Springer Series in Plasma Science and Technology

Plasma Science and Technology covers all fundamental and applied aspects of what is referred to as the "fourth state of matter." Bringing together contributions from physics, the space sciences, engineering and the applied sciences, the topics covered range from the fundamental properties of plasma to its broad spectrum of applications in industry, energy technologies and healthcare.

Contributions to the book series on all aspects of plasma research and technology development are welcome. Particular emphasis in applications will be on high-temperature plasma phenomena, which are relevant to energy generation, and on low-temperature plasmas, which are used as a tool for industrial applications. This cross-disciplinary approach offers graduate-level readers as well as researchers and professionals in academia and industry vital new ideas and techniques for plasma applications.

More information about this series at http://www.springer.com/series/15614

Tetsuo Tanabe

Plasma-Material Interactions in a Controlled Fusion Reactor

Springer

Tetsuo Tanabe
Research Center for Artificial Photosynthesis
Osaka City University
Osaka, Japan

Kyushu University
Fukuoka, Japan

ISSN 2511-2007 ISSN 2511-2015 (electronic)
Springer Series in Plasma Science and Technology
ISBN 978-981-16-0327-3 ISBN 978-981-16-0328-0 (eBook)
https://doi.org/10.1007/978-981-16-0328-0

This Springer imprint is published by the registered company Springer Nature Singapore Pte Ltd.
The registered company address is: 152 Beach Road, #21-01/04 Gateway East, Singapore 189721, Singapore

Preface

Since the discovery that nuclear reactions produce energy, 90 years have already passed. Now fission reactors are well established as energy sources, while a fusion reactor seems to need still several decades to be realized. Even though the assembly of ITER (the International Tokamak Engineering Reactor) has started on July 28, 2020, it will not be used as an energy source. Why has such a long time been required to establish a fusion reactor as an energy source compared to fission reactors? The answer could be found in different energy conversion mechanism between the two.

In a fission reactor, the energy required to start nuclear chain reactions and to keep them in steady-state operation is quite small. Energy is used just for adjusting the position of control rods including neutron absorbers or the density of neutron absorber in cooling water to maintain steady operation and to stop the reactions by inserting control rods in the reactor core. In order to convert the output power produced by fission reactions, a massive flow of coolant, i.e. water in most operating fission reactors, should be circulated and this requires a significant amount of energy, similar to that required by normal power stations using oil or coal as a fuel and would be required by a fusion reactor as well.

Unlike the fission reactor, a huge power is required for starting and keeping the burning plasma in a fusion reactor. Accordingly, the power used is dissipated to plasma-facing surfaces (PFS). In particular, the power load to the divertor area is extremely high so that power exhaust is one of the critical issues to establish the fusion reactor. The power load to PFS appears as plasma–materials interactions (PMI), which are similar to what the surface of a rocket running into the sun would be exposed to and, at present, the physical and chemical phenomena expected in PMI in a fusion reactor would be very difficult to study directly.

Among presently operating plasma apparatus, only JET can realize similar or a little lower levels of power load. Therefore, understanding of PMI in a fusion reactor must be extrapolated from observations on the currently operating apparatus whose power load is still too low. Hence special apparatus such as linear plasma machines to realize a heat load like the divertor region of a fusion reactor are being constructed. Heat exhaust is also one of the most important technical issues in the development of a fusion reactor.

Thus, PMI is a critical scientific issue for the establishment of a fusion reactor as a power source, with major potential limitations on plasma core and edge operating parameters. Gaining understanding and predictive capability in this area will require simultaneously addressing complex and diverse physics and chemistry occurring over a wide range of lengths (angstroms to meters) and times (femtoseconds to days). Furthermore, the following key engineering issues remain: lifetime of plasma-facing components (PFC's) due to steady-state sputter erosion; erosion/damage by plasma transients; surface ultrastructure and mixed-material evolution; plasma contamination by eroded material and plasma operating limits due to these factors. Shortage of tritium fuel resources and radioactivity give additional issues for fuel self-sufficiency and nuclear safety in controlling PMI.

Although many books and reviews have already been published on PMI some of which are referred in Chap. 1, they are mostly standing on plasma physics, namely how plasma or plasma confinement is influenced by PMI. This book stands on the material side focusing on changes caused by heat and particle load, i.e. how plasma-facing materials (PFM) are modified by plasma exposure and then accordingly how the modified PFM responds to the plasmas.

After the introduction of PMI in Chap. 1, Chap. 2 describes what present tokamak discharges look like and Chap. 3 describes the power load on PFM. Then basic processes of PMI are described in Chaps. 4–7: "Responses of plasma-facing surfaces to heat and particle loads" in Chap. 4, "Erosion and deposition and their influence on plasma behavior" in Chap. 5, "Material modification by high-power load and its influence on plasma" in Chap. 6, and "Fundamentals of hydrogen recycling and retention" in Chap. 7. In Chap. 8, "PMI in large Tokamaks" is discussed and in Chap. 9, "Estimation of tritium retention in a reactor" is discussed.

In Chap. 10, "Selection of PFM" is discussed. Finally in Chap. 11, "Closing remarks, including future prospect" is given.

Hopefully, this book will help in the understanding of PMI and its impact on plasma confinement and will preview what kind of research and development will be required.

Osaka, Japan Tetsuo Tanabe

Acknowledgements

The author greatly appreciates the work of graduate students in his laboratory; Drs. Kaname Kizu, Kei Masaki, Takahiro Shibahara, Kazuyoshi Sugiyama, Kazutaka Miyasaka, and Masashi Yoshida. Without their efforts, this book could not have been prepared. Many works referred here are based on TEXTOR cooperation under the framework of IEA (the International Energy Agency) and collaborations with Drs. E. Vitzke, V. Phlipps, M. Rubel, A, Pospesczyck, S. Bresinsek, G. Sergienko, A. Kirshcner, A. Huber, M. Rubel, J. P. Coad, N. Bekris, T. Hirai, T. Ohgo, Y. Gotoh, N. Noda, M. Wada, and Y. Sakawa are highly appreciated. JAERI-University cooperative researches on PWI in JT-60U with Drs. N. Miya, Y. Hirohata, K. Ohya, Y. Oya, T. Nakano, and Y. Ueda are also acknowledged.

* Throughout this book, the term "hydrogen" is used as the representative of all hydrogen isotopes including protium (H), deuterium (D), and tritium (T), if not specified.

Contents

Part I Fusion Reactor and Plasma Material Interactions

1 Introduction ... 3
1.1 The Organization of This Book ... 3
1.2 Plasma–Material Interactions Caused by Power Load of Radiation and Energetic Particles ... 5
1.3 Energy Conversion from Nuclear to Thermal for Electric Power Generation ... 7
1.4 Brief History of the Development of Plasma-Facing Materials ... 10
1.5 On PMI Studies for a Fusion Reactor ... 14
References ... 15

2 Discharges in Current Large Tokamaks ... 17
2.1 Introduction ... 17
2.2 Discharges of Current Large Tokamaks ... 20
2.3 Diagnostics for PMI Research ... 21
2.3.1 Optical Spectroscopy ... 23
2.3.2 Probe Measurements ... 26
2.4 PMI Observed by Proves and Limiter Experiments ... 29
References ... 34

3 Power Load on Plasma-Facing Materials ... 37
3.1 Introduction ... 37
3.2 Estimation of Power Load and Its Distribution in a Fusion Reactor ... 38
3.3 Steady-State Power Load ... 41
3.4 Transient Power Load ... 42
3.5 Power Load by Neutrons ... 44
3.6 Mitigation of Power Load (Power Exhaust) ... 45
References ... 46

Part II Basic Processes in PMI

4 Responses of Plasma-Facing Surface to Power Load Given by Radiation and Energetic Particles 49
4.1 Introduction 49
4.2 Energy Loss Processes of Energetic Particles Injected in a Solid Target 51
4.3 Emission of Ions and Neutrals 53
4.3.1 Reflection 53
4.3.2 Physical Sputtering 56
4.3.3 Chemical Sputtering 58
4.3.4 Ion-induced Desorption and Radiation-Enhanced Sublimation 61
4.4 Emission of Electrons and Photons 61
4.4.1 Electron Emission 61
4.4.2 Photon Emission 62
4.5 Energy Reflection 63
4.6 Reemission of Incident Ions 63
4.6.1 Reemission of Hydrogen (Fuel) 64
4.6.2 Reemission of Inert Gas Atoms 66
4.7 Interaction of Released Particles with Photons and Electrons in Boundary Plasmas 68
4.8 Summary 71
References 71

5 Erosion and Deposition, and Their Influences on Plasma Behavior (Material Transport in Tokamak) 75
5.1 Introduction 75
5.2 Erosion, Transport, and Deposition 77
5.3 Formation of Deposited Layers Made of Eroded Materials 79
5.3.1 Carbon Wall 80
5.3.2 Metallic Wall 91
5.4 Summary 92
References 93

6 Material Modification by High-Power Load and Its Influence on Plasma 95
6.1 Power Load to PFM 95
6.2 Material Response to Power Load and Its Influences on Boundary Plasmas 96
6.2.1 Spontaneous Response to Power Load 96
6.2.2 Melting and Sublimation 97
6.2.3 Hydrogen Recycling 99
6.3 Damaging and Degradation of PFM 99
6.3.1 Carbon (C) 100
6.3.2 Tungsten (W) 102

6.3.3 Other PFM Candidates (Be and Li) ... 109
6.3.4 Structure Materials ... 109
6.4 Summary ... 110
References ... 111

7 Fundamentals of Hydrogen Recycling ... 115
7.1 Introduction ... 115
7.2 Overall Fuel Flow at Steady-State Burning ... 116
7.3 Injection of Energetic Hydrogen ... 118
7.4 Reflection, Reemission, and Retention ... 120
7.5 Permeation ... 123
7.6 Isotope Effects ... 125
7.7 Long-Term Retention and Trapping ... 126
7.8 Simulation and Modeling ... 128
7.9 Summary ... 128
References ... 129

Part III PMI, Observations in Present Large Tokamaks and Prospects in a Reactor

8 PMI in Large Tokamaks ... 133
8.1 Power Load ... 133
8.1.1 Power Load in JET ... 133
8.1.2 Exchange of PFM from Carbon to High Z Metals ... 136
8.1.3 ITER-Like Wall (ILW) in JET ... 136
8.1.4 Power Load by High Energy Particles Produced by Fusion ... 138
8.2 Erosion and Deposition ... 138
8.2.1 Carbon Wall (C-Wall) ... 138
8.2.2 Metallic Wall ... 141
8.3 Dust ... 143
8.4 Recycling and Retention of Fuels ... 145
8.4.1 Consideration of Fuel Retention Rate ... 146
8.4.2 Recycling ... 147
8.4.3 Long Term Fuel Retention ... 152
8.5 T-Related Issues on the In-Vessel T Inventory ... 156
8.6 Summary ... 157
References ... 157

9 Fuel Retention in a Reactor with Full C-Wall and Full W-Wall and Its Recovery ... 161
9.1 Introduction ... 161
9.2 Present Estimation of Fuel Retention in ITER ... 162
9.3 Construction of Fuel Retention Model in a Fusion Reactor ... 163
9.4 Fuel Retention in Carbon Materials ... 164

9.4.1 Characteristics of Hydrogen Retention in Carbon Materials [11] 164
9.4.2 Fuel Retention Build-Up in JT-60U, a Full Carbon Wall Tokamak 167
9.4.3 Estimation of Carbon Deposition and Fuel Retention in an ITER Scale Full Carbon Reactor Operated at Around 600 K 171
9.5 Fuel Retention in Tungsten (W) 173
9.5.1 Characteristics of Hydrogen in W 174
9.5.2 Fluence Dependence of H Retention in W 177
9.6 Comparison of Estimated Fuel Retention in a Reactor with Full C-Wall and W-Wall 179
9.7 Fuel Removal/Recovery 180
9.7.1 Removal/Recovery of T Retained in Carbon Materials 181
9.7.2 Removal/Recovery of T Retained in W 182
9.8 Summary 183
References 183

10 Selection of Plasma-Facing Materials 187
10.1 Criteria for Selection of PFM 187
10.2 Concerns on W Usage as PFM 189
10.3 Use of Carbon Materials as PFM 190
10.3.1 Character of C as PFM 190
10.3.2 Possible Use of C as PFM in a Reactor 191
10.4 Liquid PFM 194
10.5 Consideration of T Fuel on the Selection of PFM in a Reactor . . 195
10.6 Summary 196
References 197

11 Closing Remarks 199

Index 201

Part I
Fusion Reactor and Plasma Material Interactions

Chapter 1
Introduction

1.1 The Organization of This Book

There is one significant difference between plasma–material interactions (PMI) and other interactions between gas and solid or liquid, and liquid and solid. For the latter cases, one can think of chemical or thermo-dynamical equilibrium where no energy transfer occurs, while in PMI there is always energy or power flow from plasma to materials under a steep gradient of (potential) energy. The higher energy state of plasma than that of plasma-facing materials causes energy release or power flow from the plasma to the materials by radiation and kinetic energy of plasma particles (ions of fuels, He and impurities, and electrons motivated by the energy gradient). Such power release by the radiation and the kinetic energy of the particles results in PMI which includes so many different physical and chemical phenomena that it is hardly possible to introduce PMI phenomena in an orderly sequence but to introduce important subjects separately. (Please refer Fig. 1.3, which describes various physical and chemical phenomena occurring in a wide range of energy states together with energy ranges corresponding to fusion reaction, burning plasma, plasma–surface interactions, material responses to the power exhaust in a fusion reactor).

In addition, fuel losses by burning and flow-out from fusion plasma must be compensated by fueling, which is another important subject of PMI in a reactor. Different from most of the plasma apparatus, tritium (T) is used as a fuel of the reactor. Since T is hazardous due to its radioactivity and its resources are scarce, special care is required for safety handling and fuel self-sufficiency. Fuel recycling at plasma-facing surface (PFS) has often been discussed separately with the power load. However, as described above, the power is carried by fuel particles. Hence, the fuel recycling should be discussed considering the power flow.

This book tried to describe PMI phenomena referring basic physics and chemistry in them and their roles in the construction of a fusion reactor. To realize this, the total of 11 chapters are grouped into three parts which correlate with each other as shown in Fig. 1.1. Part I consists of three chapters. Following the present chapter (Chap. 1) which describes the concept of PMI, brief introduction of a D-T fusion

T. Tanabe, *Plasma-Material Interactions in a Controlled Fusion Reactor*, Springer Series in Plasma Science and Technology, https://doi.org/10.1007/978-981-16-0328-0_1

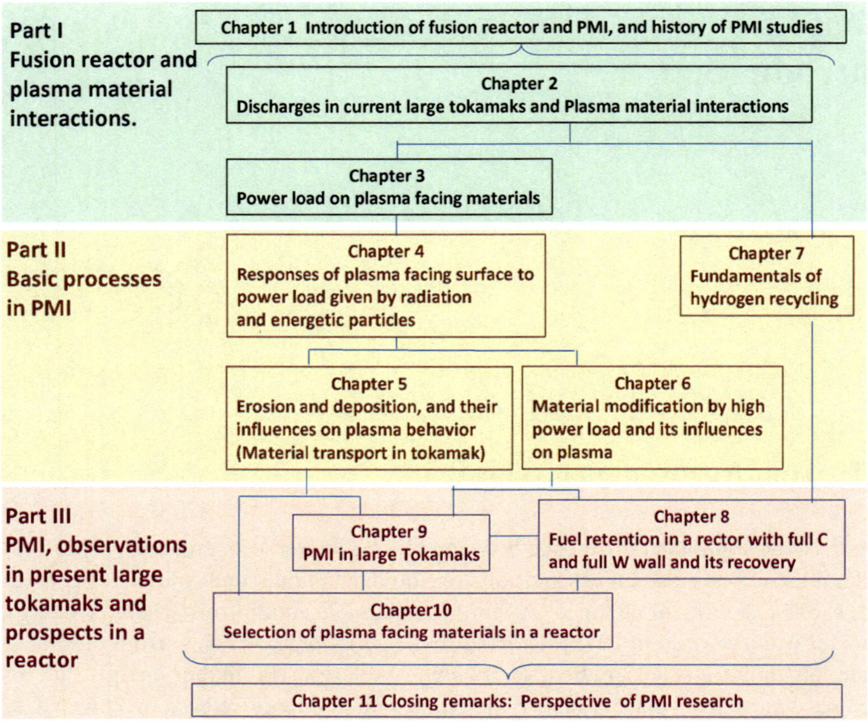

Fig. 1.1 The organization of this book

reactor and history of PMI studies, and required PMI studies in future. Chapter 2 describes the observation of discharges in a large tokamak, and Chap. 3 explains power load to plasma-facing surface as the main cause of PMI in a fusion reactor. In Part II, four basic processes in PMI are separately explained. They are "Responses of plasma-facing surface to power load" in Chap. 4, "Erosion and deposition, and their influences on plasma behavior" in Chap. 5, "Material modification by high-power load and its influences on plasma" in Chap. 6, and "Fundamentals of hydrogen (fuel) recycling" in Chap. 7. In Part III is described the present research status and future prospect of PMI in three chapters. Chapter 8 summarizes "PMI in large Tokamaks". Chapter 9 describes "Fuel retention in a rector with full C and full W wall and its recovery" as one of the critical issues for the establishment of a fusion reactor as an energy source. Chapter 10 gives some ideas on the "Selection of plasma-facing materials" for which we do not have the solution yet. In Chap. 11, the final chapter, future prospects and research targets of PMI are suggested as the closing remarks.

1.2 Plasma–Material Interactions Caused by Power Load of Radiation and Energetic Particles

Nearly 90 years have passed after finding that nuclear reactions give energy. Now fission reactors are well established as energy sources, while a fusion reactor seems to need still a few decades to be realized. Why so much longer time has been required to establish a fusion reactor as an energy source compared to fission reactors?

There have been various difficulties in the research and development of a fusion reactor as an energy source. Although confinement of burning plasma has been the largest hurdle, it could be overcome in ITER. In engineering aspects, several problems are coming up, such as extremely high-power load to plasma-facing materials (PFM), conversion of neutron energy to heat, and management of radioactive T fuel to keep safety and fuel self-sufficiency. For consideration of the power load and energy conversion, principles of energy conversion in a fusion reactor system and a fission reactor system are quite different as compared in Table 1.1. Different from any other energy sources, the fusion reactor needs a significant amount of energy to start burning or ignition and also to continue burning, i.e. to make high-energy and high-density

Table 1.1 Comparison of energy conversion processes in fission and fusion reactors as energy sources

	Fission reactor	Fusion reactor
Characteristics	All of the energy conversion, fuel breeding, waste-confinement are done in a fuel pin of diameter of ~1 cm	An open tritium handling system with a huge volume
Power Input	Nearly zero	Huge power is required to sustain and to keep burning plasma Poor fueling efficiency requires huge fuel throughput
Energy conversion	Energy carried by fission products (FP, heavy ions) (~170 MeV) is deposited in fuel pins and converted to heat	Energy carried by a neutron (14 MeV) must be converted to heat in a large volume of the blanket system
Fuel breeding and recovery	One fission produces more than two neutrons, easy to keep chain reactions and to breed fuels Fuel pins contain both FPs and new fissile Spent fuels are reprocessed to remove/recover them	To keep breeding ratio more than 1, neutron multipliers (Be, Pb) are required Tritium breeding and energy conversion must be done simultaneously
Nuclear Waste	Long-life radioactive FPs and trans-uranium elements must be handled with special care and will be reposed deeply underground	Waste is limited to activated structure materials and could be recycled

plasma confined to satisfy the Lawson condition. The initial input power would be 1/3 to 1/4 of the fusion output power. Since a fusion reactor will be designed to produce the power of a few GW, each reactor may require a power station with the power of a few hundred MW to start up. In a fission reactor, no such high power is required, and only removing control rods from the reactor core starts fission reactions and continue steady-state burning.

Energy conversion systems for fission and fusion are also completely different. In a fission reactor, the energy produced by the fission reaction of an uranium atom (U) and a neutron is carried mainly by fission products (FPs) and transformed to thermal heat of coolant for electric power generation, while, in a fusion reactor, the energy carried by 14 MeV neutrons must be converted to the thermal heat of the coolant in the blanket. At the same time, the 14 MeV neutron is used to breed T to sustain fuel self-sufficiency as described below. Both fission and fusion leave nuclear wastes. Compared to long-life nuclear wastes in fission including FPs and trans-uranium elements like U, Np, and Pu, which are serious concerns for radiation safety, activated structure materials in a fusion reactor by neutron irradiation are less hazardous and even they could be re-used.

Any power sources require power removal through cooling system to convert their generated energy to heat or electricity. In a fusion reactor, nearly 1/4 of generated fusion power is used to sustain burning plasma (either initial heating power from outside and heating power given by He) and must be removed or recovered. Except energy carried by the 14 MeV neutrons, all power used for plasma heating is loaded to plasma-facing materials (PFM). Divertor is introduced to remove such high-power load and He ash. Still the tolerance of PFM to the power load is concerned. PFS of the main chamber is also exposed to the significant power load. Depending on the location of the plasma-facing components, power loads significantly differ. Mid-plane of the central pole (inner first wall) would be the highest except divertor target area.

Power load is given to PFM by radiation or energetic photons, and energetic particles including tritons, deuterons, neutrons, helium ions, and electrons escaping from boundary plasma in addition to neutrals produced by charge exchange. The radiation consists of the Bremsstrahlung emission from burning plasma, and radiation from impurities in plasma and seeded gas required for cooling the boundary plasma. The power load to PFM is so high that no simple material can tolerate without active cooling. Still there is a limit in the removal of the loaded power. Accordingly, the maximum of the power load to materials having a high melting point is limited to 10–20 MW m^{-2} under efficient cooling.

The main subjects of this book entitled "Plasma wall interactions in a fusion reactor" are to introduce/discuss responses of PFS to the power load and modification of PFM by the power load, both of which significantly influence the performance of burning plasma.

In the early days of PMI studies, when the first proceedings of the International Symposium on Plasma Wall Interaction were published in 1977 [1], the main interest was focused on material response to the injection of high-energy ions appearing as sputtering, and little works had done on high-power load, because confined energy in the plasma was too small to observe the effect of the heat load on PFS at that time. Since then, many books on PMI or plasma–surface interactions (PSI)* have been published. Some are given in references mostly focusing on sputtering and some atomic processes of sputtered atoms in boundary plasma [2–15]. Based on these PMI studies, plasma technologies are now widely used for the production of new materials and/or material modification [16], which are not discussed here.

* Plasma–material interaction (PMI) or plasma–surface interaction (PSI) are often used to represent the same meaning, so as plasma-facing surface (PFS) and plasma-facing materials (PFM). In this book, generally PMI and PFM are used, while PSI is used to represent phenomena in which only PFS is involved.

1.3 Energy Conversion from Nuclear to Thermal for Electric Power Generation

Figure 1.2 shows how energy released by a DT fusion reaction is converted to thermal energy. The D-T reaction produces 14 MeV neutron and 3.5 MeV He.

$$D + T \rightarrow {}^{4}He(3.5MeV) + n(14.1MeV) \tag{1.1}$$

The main process to get electric power is energy conversion of the neutron energy (14 MeV) to thermal heat of coolant (Rotational and vibrational motions of atoms or molecules consisting of coolant or around 0.1 eV phonon) in the blanket. The 14 MeV neutron directly goes into the blanket and its energy is first lost by a nuclear collision

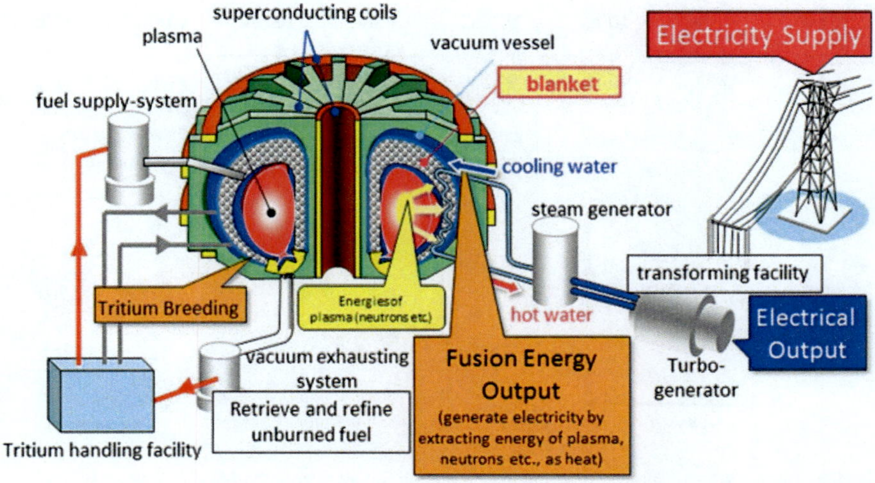

Fig. 1.2 Fusion reactor system and energy conversion. http://www.fusion.qst.go.jp/rokkasyo/en/project/blanket.html

with atoms consisting of the blanket. Neutron collision with Li in the blanket produces tritium (T). Both recoil atoms and produced T are in high-energy states (MeV to keV range). Succeedingly, they lose their energy with either electric or nuclear collision processes, and finally converted to heat (thermal energy) that is carried by vibration and rotation of atoms and molecules and collective motions of atoms consisting of the blanket, i.e. phonons with the energy of several tens meV (several hundred Kelvin). During the collision processes, secondary electrons and photons are produced by ionization and relaxation of atoms. The secondary particles lose their energy in a similar way until particle energy becomes smaller than the energy required for ionization or electron excitation. Most of these energy conversions are done in the blanket. According to the figure, the energy of 14 MeV neutron is converted into thermal energy consisting of around 10^8 phonons with an energy of meV range less than millisecond. The thermal energy is transferred to coolant for the generation of electricity through turbines. Understanding of the energy conversion processes will require simultaneous addressing of complex and diverse physics and chemistry occurring over a wide range of energy (MeV to meV), lengths (angstroms to meters), and times (femtoseconds to days) as shown in Fig. 1.3. In the figure are also given energy ranges corresponding to phenomena occurring in fusion reaction, burning plasma, boundary plasma, plasma-facing surface, and plasma-facing materials.

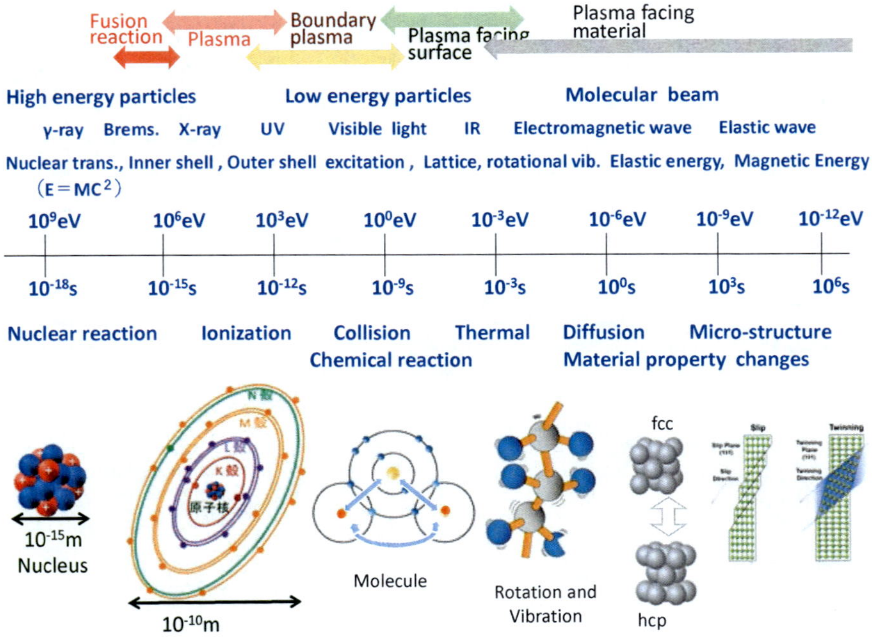

Fig. 1.3 Energy transfer/conversion and accompanied physical and chemical processes with characteristic times and their scales or sizes. Required energies to promote these processes correlate the times and sizes of the processes

There is one more important energy conversion process, which occurs at PFS. For sustaining burning plasma, fusion plasma shall be confined in the magnetic field. Confinement times of fuels and energy in the burning plasma are quite short, only a few seconds or less. That means fuels and energy are supplied and exhausted simultaneously to keep the input and output balance of fuel and power, which are quite hard to realize. The fuel balance has been discussed in a separate book [17] and will not be discussed here. To start up the burning plasma, energy is supplied from outside by various techniques, such as ohmic heating, neutral beam injection, electron cyclotron heating, ion cyclotron heating, and so on, while at the steady state, self-heating by 3.7 MeV He is expected in addition to the ohmic heating. Since plasma temperature is around 10–20 keV, 3.7 MeV He collides with fuel ions and electrons in plasma to give its energy to them through either the Coulomb collision or nuclear collisions. If the burning plasma does not contain impurity other than He, the Bremsstrahlung photon emission caused by motions of electrons, fuel ions, and He ions in the magnetic field gives energy loss from the plasma core.

Real plasma includes various impurities, in particular, atoms consisting of plasma-facing materials, and impurities included in them are easily released by collisions with plasma particles escaping from the plasma. Once these impurities enter in plasma, they are immediately ionized. Then they capture electrons and emit photons with energy corresponding to energy levels of binding electrons in impurity atoms. Together with the Bremsstrahlung emission, radiation from the impurities reduces plasma temperature. Depending on their atomic number, radiation power when they are in plasma changes significantly as shown in Fig. 1.4 [18]. It should be noted that depending on radiated energy which is proportional to the frequency or inversely proportional to the wavelength of radiated photons, the character of radiation or its

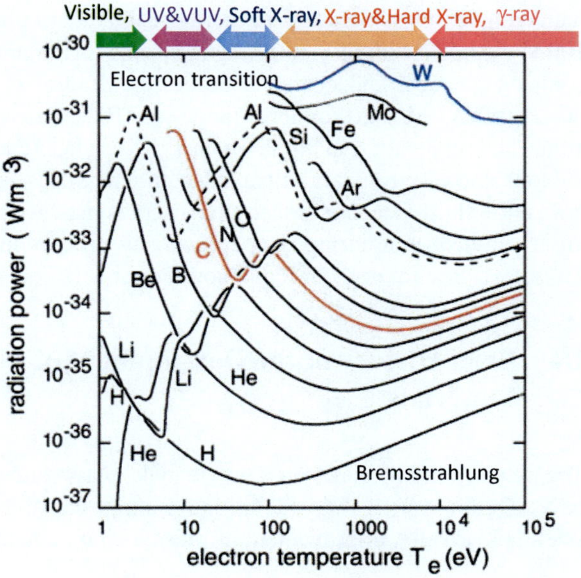

Fig. 1.4 Plasma (electron) temperature dependence of radiation power for various elements based on the Corona model. (reprinted with permission from [18]) On the upper horizontal scale, names of radiation are indicated

interactions with materials are different. Photons in the energy range of 1 eV to several eV are visible, 10 eV to several tens change from ultraviolet (UV) to vacuum ultraviolet (VUV), and above 100 eV changes from Soft X-rays to keV range X-rays as indicated in the figure. This means that depending on the atomic numbers of the elements, not only their radiation power but also the character of their radiation is different. Higher Z elements like Tungsten (W) are most probably used as PFM in a fusion reactor. When they are accumulated in the plasma center, they give stronger radiation as the X-rays and hence their concentration in core plasma should be suppressed to be below 10^{-5} to 10^{-6}.

Interactions of VUV and Soft X-ray with materials are so intense that they can penetrate the materials only in very short distances, in other words, their energy deposition on the material is so intense to exhibit strong PSI. Since the radiation for the high Z elements is in the X-ray region, PSI for the high Z wall must be dominated with the interaction with strong X-ray. However, no such high-intensity sources of Soft X-ray and X-rays emitted from burning plasma are available, the interaction of the strong radiation from the accumulated high Z elements in the plasma center with materials is not easy to study and lots of subjects remain, for examples, plasma opacity (radiation from plasma center does not come out to PFS to inhibit power exhaust), the effect of energy deposition limited within very thin surface layers, vapor shielding, and so on..

Particle confinement time in plasma is limited within a few seconds and they escape from the plasma with diffusion across the magnetic field. Since the surface temperature of the PFS should be below their melting temperature, injection of high-energy plasma particles to PSF should be avoided. To realize this, boundary plasma or scrape off layers are constructed with the installation of limiters or divertor outside of the last closed magnetic field lines. The energy conversion from 10 keV in core plasma to 100 eV in boundary plasma occurs with various processes together with complex mass transfer. Furthermore, boundary plasma with the energy of 10–100 eV contacts with PFS, resulting in plasma–material interactions (PMI) including complex physical and chemical processes. Studies of the physical and chemical processes occurring in the energy range of 1–100 eV are quite difficult mainly because atoms and molecules having these energies are in excited states. Moreover, emitted photons and electrons accompanied with energy loss process in this energy range have quite short escaping (penetration) depths in materials even in gases. This makes observation or monitoring of the occurring process in boundary plasma difficult. Thus, PMI has not been well diagnosed or understood yet.

1.4 Brief History of the Development of Plasma-Facing Materials

The first step in fusion research was to make plasma in vacuum, or to confine plasma in a vacuum system that was initially composed of glass and turned to stainless steel. Historically, improvement of plasma confinement relied on the improvement of vacuum and the control of impurities in plasma. In the early stage of plasma researches, glass was used as a vacuum vessel because of its small desorption of impurities, in particular, water. With the improvement of vacuum pumps, metals have replaced the glass. Improvement of plasma confinement has proceeded with the development of vacuum technology. Nevertheless, radiation from impurities in plasma, mostly carbon and oxygen originated from residual gas in a vacuum, prohibited raising plasma temperature. In some sense, an increase in plasma temperature or improvement of plasma confinement is fighting with radiation loss caused by impurities. From the beginning, plasma researchers were targeting to develop fusion reactors as an energy source. Hence, high melting temperature materials and/or materials admitting cooling came in sight as PFM. Various materials were tested, starting from the glass, stainless steel, Cu for easy cooling, carbon with no melting, and high melting temperature metals like Mo and W. Even liquid Li has been used. For any materials used as a vacuum wall, radiation or power load induces desorption of water and hydrocarbons adsorbed on the wall to retard making a good vacuum. Hence removing water and hydrocarbons was the key to improve the vacuum and plasma confinement. In addition, utilization of W as PFM in ORMAK plasma in 1979 [19] gave significant central radiation as shown in Fig. 1.5 mainly because of W accumulation in the plasms center. This was the first clear observation of plasma contamination by the wall material. Since then, W or other high Z materials had been excluded to use as PFM until TEXTOR employed Mo limiter in 1993 [20–22]. (Nowadays as discussed later, concerns of T retention and neutron damage of carbon forced to use W as PFM of ITER).

Therefore, the reduction of impurities in plasma was the first priority to get good plasma performance. From 1984 to 1990, every two years when an "international conference on plasma–surface interactions in controlled fusion" was held, significant development of vacuum technique was presented, accompanying the improved

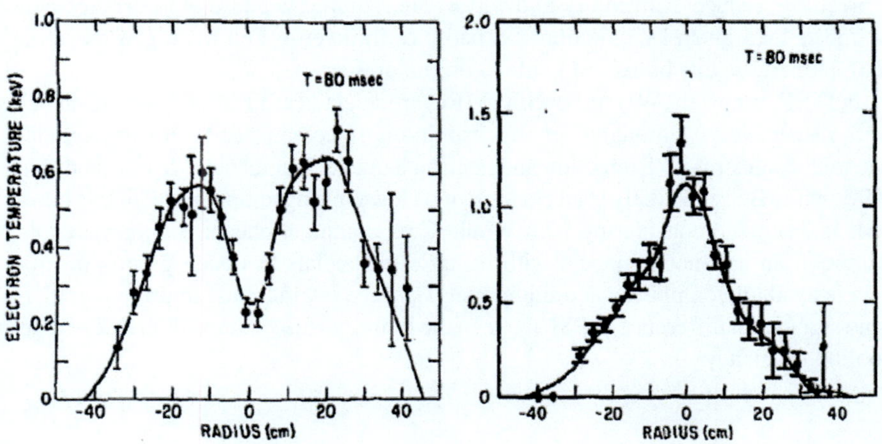

Fig. 1.5 **a** Hollow temperature profile caused by impurity accumulation in plasma and **b** Recovery of central peaked profile in ORMAK tokamak (reprinted with permission from [19])

plasma confinement as shown in Fig. 1.6. The first break in the early '80s was given by titanium (Ti) gettering to give a good vacuum and hence confinement was significantly improved. Covering all PFS by deposited carbon layers (Carbonization) [22] further improved plasma performance, and plasma temperature over 1 keV was attained. With this, in the late '90s, the temperature barrier owing to strong radiation of oxygen around 600 eV to inhibit plasma heating was overpassed. Introduction of boron (B) [23] and beryllium (Be) [24, 25] which capture oxygen and radiate less results in a significant increase in plasma temperature. As shown in Fig. 1.7 [26], the advance of plasma confinement in the 80's is quite parallel to these first wall changes.

When plasma density and temperature became sufficiently high, a new problem appeared; the difficulty of density control owing to reemission of hydrogen (recycled hydrogen) from PFS and temperature rise of PFM. In recent days, for easy density control and good plasma performance, wall pumping is encouraged, which means PFS retains most of the injected hydrogen during a discharge with less recycling of hydrogen. To attain wall pumping conditions, wall cleaning or hydrogen removal from the wall by various methods was employed. Today discharge cleaning, which uses glow or ECH discharges with an inert gas like He as working gas, is routinely used in most tokamaks. With these successes, plasma temperature has raised over several keV in three large tokamaks in the world, JET, TFTR, and JT-60U and burning plasma looked like within our hands. On the other hand, heat load to PFM in such high-temperature plasma became so large to enhance evaporation and melting of PFM.

To reduce heat load to PFS and to exhaust He, a divertor structure was introduced. To realize a fusion reactor as an energy source, one of the hardest tasks is to establish a divertor system that tolerates high-power load (remove heat load) and exhausts He quite effectively. Since T fuel of a fusion reactor is radioactive and its resources are limited, keeping T safety in construction and operation and having enough margin in T breeding (reduce T inventory in divertor) are required in addition to management of power load given by radiation and particles in divertor and hydrogen recycling [26]. Hopefully, this book will guide to or hint at that.

In ITER, tungsten (W) and beryllium (Be) are selected as PFM of divertor and first wall, respectively. Although Carbon was initially selected as PFM, it was excluded due to the concerns of T retention and neutron damage. In a reactor, W is a candidate PFM, while Be is not likely used because of its low melting temperature. In this book, PMI is discussed considering both W and C as candidate plasma-facing materials. Although an advanced concept with using liquid metals has been proposed, little data is available on plasma and liquid metal surface interactions. In principle, there is no significant difference in PSI between the liquid surface and solid surface, except cooling efficiency.

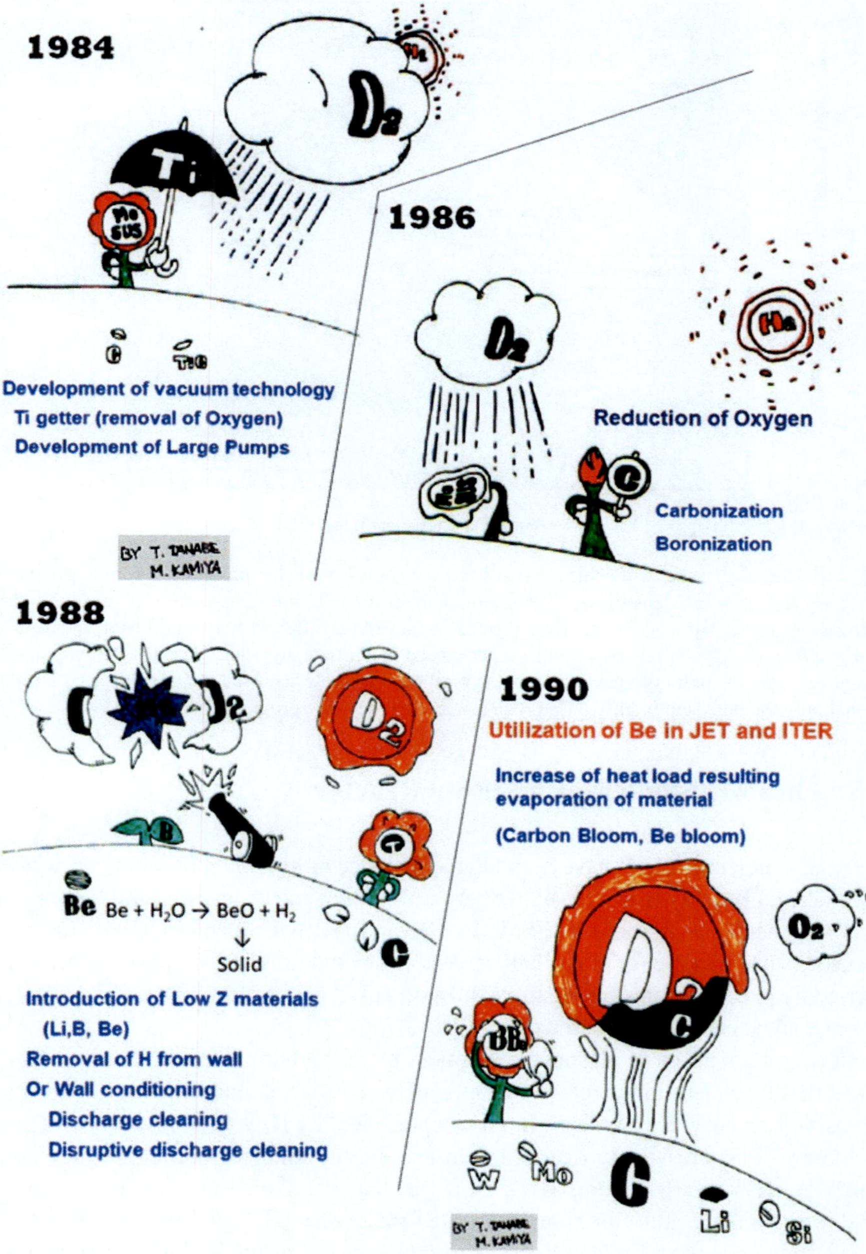

Fig. 1.6 Historical changes of plasma-facing materials from 1984 to 1990

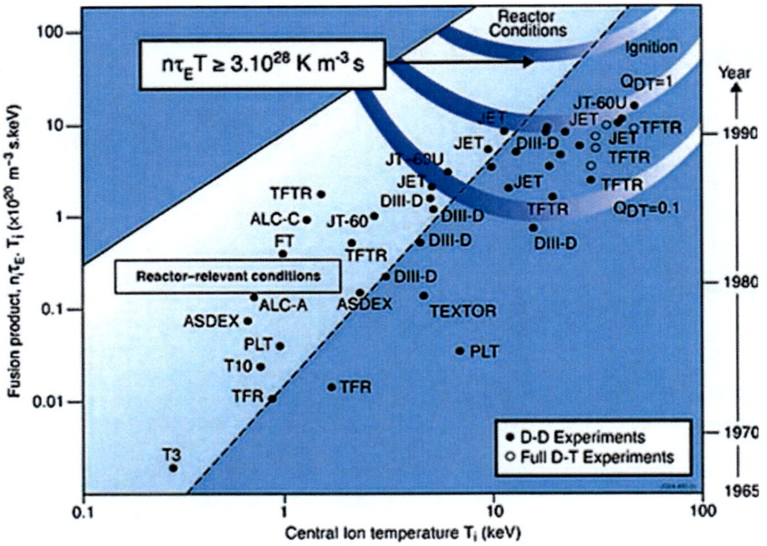

Fig. 1.7 Plasma confinement achieved on different fusion facilities, represented by triple product ($n_i \times \tau_E \times T_i$); n_i ion density, τ_E energy confinement time, T_i ion temperature. (reprinted with permission from [26]) The fusion triple product achieved on different magnetic fusion facilities. The graph shows clearly that new facilities performed better than previous ones. The present large machines, from the point of view of the fusion product, have now achieved their engineering limits so that only the next step facility, ITER, can bring about decisive progress

1.5 On PMI Studies for a Fusion Reactor

The main subjects of PMI have been (1) fuel balance or hydrogen recycling between plasma and PFM, (2) material balance between erosion and deposition, and (3) power load (or energy deposition) to PFM. In conventional hydrogen recycling studies, focused is just particle (or mass) balance, but little attention has been paid on energy carried by particles contributing to PMI. In other words, particle balance and power balance are separately discussed.

In recent advances of plasma confinement targeting burning plasma, it is recognized that consideration of power balance between plasma and PFM or net power deposition on PFM is necessary. In fuel (represented by H) recycling, fuel particles incident to PFS are widely distributed in energy and kinds of particles which are ions (H^+, H^-), excited atoms (H^*), atoms in the grand state (H), molecular ions ($H_2{}^+$), rotationally and vibrationally excited molecules (H_2^*), and molecules in the grand state (H_2), so as released fuel particles are. Accordingly, incident particles give power to PFM and released particles remove power from PFM. The incident fuel particles induce emissions of particles as sputtering, ion-induced desorption, electrons, photons, and phonons and remain as radiation damages in PFM. All these particles and photons emission remove power from PFM. The net deposited power to PFM is the balance among these three and contributing PFM temperature rise of which increment is depending on cooling power, and influences hydrogen recycling. In addition, some impurities like C and O always exist in plasma and sometimes largely contribute to PMI. In particular, radiation from impurities in plasma exceed radiation from fells and often disturbs power balance.

Studies of fuel recycling between plasma and PMI are not enough. Fuel recycling, i.e. throughput, exhaust, in vessel retention, breeding and recovering in blanket, and refuelling, is one of the keys of a fusion reactor. For T safety and fuel self-sufficiency, fuel balance among throughput, exhaust, and retention in reactor vessel should be precisely measured. Nevertheless, the balance has not been considered seriously in the present tokamaks. Fuel retention in the reactor vessel involves not only retention in PFM but also those in deposits at non-plasma-facing surface and remote area resulting from PMI. Therefore, in this book, fuel retention is explained in a little detail.

Although various simulation techniques are advancing and used for understanding PMI including fuel recycling, such complex PMI processes, i.e. particle transport accompanying energy or power changes, are quite hard to simulate with conventional methods except particle simulations. At present, full particle simulation including all PMI processes is not possible. Still experimental works are quite important with the consideration that particle transport always accompanies energy or power transport. Hopefully, this book encourages readers to perform PMI studies necessary to establish a fusion reactor as an energy source.

References

1. S. Stuart, *Proceedings of the International Symposium on Plasma Wall Interaction*. Pergamon (1977). eBook ISBN: 9781483136202
2. K. Kaminsky, Radiation effects on Solid surfaces, Ed. M. Kaminsky, American Chemical Soc., Washington (1976). ISBN: 0-8412-0331-8
3. P.D. Townsend, J.C. Kelly, N.E.W. Hartley, *Ion Implantation, Sputtering and their Applications*. Academic Press (1976). ISBN:0-12-696950-7
4. R. Behrisch, *Sputtering by Particle Bombardment I, Physical Sputtering of Single-Element Solids*. Springer (1981). https://doi.org/10.1007/3-540-10521-2
5. R. Behrisch, *Sputtering by Particle Bombardment II, Sputtering of Alloys and Compounds, Electron and Neutron Sputtering, Surface Topography*. Springer (1983). https://doi.org/10.1007/3-540-12593-0
6. R. Behrisch, K. Wittmaack, *Sputtering by Particle Bombardment III, Characteristics of Sputtered Particles, Technical Applications*. Springer (1991). https://doi.org/10.1007/3-540-53428-8
7. R.K. Janef, H.W. Drawin, *Atomic and plasma-Material Interaction Processes in Controlled Thermonuclear Fusion*, Ed. Elsevier (1993). ISBN: 0-444-81630-5
8. W.O. Hofer, J. Roth, *Physical Processes of the Interaction of Fusion Plasmas with Solids*. Academic Press (1996). ISBN: 0-12-351530-0
9. M. Nastasi, N. Michael, J Mayer, et al., *Ion-Solid Interactions: Fundamentals and Applications*. Cambridge University Press (1996). https://doi.org/10.1017/CBO9780511565007
10. D. Naujoks, *Plasma-Materials Interactions in Controlled Fusion, Springer Series on Atomic, Optical, and Plasma Physics*, vol. 39 (Springer, Berlin Heidelberg, 2006)

11. W.O. Hofer, J. Roth, *Physical Processes of the Interaction of Fusion Plasmas with Solids (Plasma-Materials Interactions).* Academic Press (February 28, 1996)
12. R.E.H. Clark, D.H. Reiter, *Nuclear Fusion Research Understanding Plasma-Surface Interactions*, Springer Series in Chemical Physics, Vol. 78, 2005, Springer, Berlin, Heidelberg. ISSN: 0172-6218
13. R. Behrisch, W. Eckstein, *Sputtering by Particle Bombardment: Experiments and Computer Calculations from Threshold to MeV Energies*, Springer Science & Business Media (2007)
14. D.E. Post, R. Behrisch, *Physics of Plasma-Wall Interactions in Controlled Fusion.* Springer Science & Business Media (2013)
15. R.E. Johnson, *Energetic Charged-Particle Interactions with Atmospheres and Surfaces.* Springer Science & Business Media (2013)
16. I. Adamovich, S.D. Baalrud, A. Bogaerts, et al. The 2017 Plasma roadmap: low temperature plasma science and technology. J. Phys. D: Appl. Phys. **50**, 323001 (46 pp) (2017)
17. T. Tanabe, *Tritium: Fuel of Fusion Reactors.* Springer Japan (2017). https://doi.org/10.1007/978-4-431-56460-7
18. R. Schneider, X. Bonnin, K. Borrass, et al. Plasma edge physics with b2-eirene. Contributions Plasma Phys. **46**, 3191 (2006)
19. R.J. Hawryluk, K. Bol, N. Bretz et al., The effect of current profile evolution on plasma-limiter interaction and the energy confinement time. Nucl. Fusion **19**, 1307–1318 (1979)
20. T. Tanabe, V. Philipps, B. Unterberg et al., High Z limiter test in TEXTOR. J. Nucl. Mater. **212–215**, 1370–1378 (1994)
21. V. Philipps, T. Tanabe, Y. Ueda et al., Molybdenum test-limiter experiments in TEXTOR. Nucl. Fusion **34**, 1417–1429 (1994)
22. J. Winter, Carbonization in tokamaks. J. Nucl. Mater. **145–147**, 131–144 (1987)
23. M. Saidoh, N. Ogiwara, M. Shimada et al., Initial Boronization of JT-60U Tokamak Using Decaborane. Japanese J. Appl. Phys. **32**, 3276–3281 (1993)
24. Karl Jürgen Dietz & The Jet Team, Properties and performance of beryllium and carbon for plasma-facing components. Fusion Technol. **19**, 2021–2028 (1991)
25. R.W. Conn, R.P. Doerner, J. Won, Beryllium as the plasma-facing material in fusion energy systems—experiments, evaluation, and comparison with alternative materials. Fusion Eng. Design **37**, 481–513 (1997)
26. 50 years of Lawson criteria https://www.euro-fusion.org/news/detail/detail/News/50-years-of-lawson-criteria/

Chapter 2
Discharges in Current Large Tokamaks

2.1 Introduction

In Table 2.1 are summarized main plasma parameters of large- or medium-sized tokamaks operated, in operation, and into operation soon. The plasma–surface interaction (PSI) studied in these machines are referred in this book. Most of the tokamaks were initially operated with Carbon (C) as plasma-facing surface (PFS) and some have changed from C to Tungsten (W). Hence, the accumulation of data on plasma–surface interactions (PSI) and experiences of tokamaks operation with C-wall have made significant improvement in plasma confinement. After finding that high Z plasma-facing wall could sustain high-temperature plasma and accordingly ITER decided to use W as the plasma-facing material (PFM) of divertor, many tokamaks given in the table have employed W as PFM. Hence plasma–material interactions (PMI) data useful for its understanding are mostly given for the C-wall, and those for the W-wall are still limited.

When the temperature of plasma was not high enough to give power load to increase the temperature of plasma-facing surface (PFS), observed radiation was mostly from plasma in the wavelength region of ultraviolet to visible light. However, as described in Chap. 1, with the increasing power load to PFS, radiation from PFS became appreciable, and accordingly, PMI interactions became appreciable. In the early days, PFS was considered passive against plasma with little influence on it. Nowadays, owing to high-power load, PFS actively responds to plasma exposure giving a significant impact on it. In particular, PMI phenomena in divertor area play a significant role in plasma confinement through fuel recycling, He pumping, power exhaust, impurity production, and so on.

T. Tanabe, *Plasma-Material Interactions in a Controlled Fusion Reactor*, Springer Series in Plasma Science and Technology, https://doi.org/10.1007/978-981-16-0328-0_2

Table 2.1 Main plasma parameters of large- or medium-sized tokamaks

Name of Tokamak	Location	Major Radius (m)	Minor Radius (m)	Magnetic field (T)	Heating power (MW)					Plasma current (MA)	Configuration	Year(s) of operation	Plasma-facing material
					Total	ECH	ICH	NBI	LH				
ALCATOR C-Mod	USA, MIT	0.67	0.22	8	8		6	DNB only	2.5	2	D-shape Divertor (single or double null)	1993–2016	Mo
ASDEX-upgrade	Germany, Garching	1.65	0.5–0.8	3.9	30	4	6	20		1.4	D-shape Divertor	1991	Full Carbon ↓ W-coated C
DIII-D	USA, San Diego	1.67	0.67	2.2	31	6	5	20		3	D-shape double null divertor	1986	Full Carbon with DiMES for materials test
EAST	China, Hefei	1.75	0.43	5.0 (SC)	7	0.5	3		4	0.5	D-shape Divertor (single or double) and limiter	2006	Full Carbon ↓ W-coated C
ITER	France, Cadarache	6.2	2	5.3 (SC)		20	20(40)	33(50)	0(40)	15	D-shape Divertor	2025	Be first wall W divertor

(continued)

Table 2.1 (continued)

Name of Tokamak	Location	Major Radius (m)	Minor Radius (m)	Magnetic field (T)	Heating power (MW)					Plasma current (MA)	Configuration	Year(s) of operation	Plasma-facing material
					Total	ECH	ICH	NBI	LH				
JET	Oxfordshire UK	2.96	0.96	4.0	38		12	24	7	6.0	W-shape Divertor	1984	Full Carbon ↓ W-coated C & Be
JT-60SA	Japan, Naka	3.16	1.02	2.7(SC)	50	7		24+10		5.5	D-shape Divertor (double null)	2021	Full Carbon ↓ W-coated C (planed)
JT-60U	Japan, Naka	3.4	1.1–1.5	2.25	60	4	10	40+10	8–12	5	D-shape Divertor	1991–2010	Full Carbon
KSTAR	S Korea, Daejeon	1.8	0.5	3.5(SC)	3(24)	0.5(4)	2(6)	(14)		2	D-shape Divertor	2008–	Full Carbon ↓ W-coated C (scheduled)
TEXTOR	Germany, Juelich	1.75	0.47	2.8	9	1	4	4		0.8	Circular Limiter/ Dynamic Ergodic Divertor	1963–2013	Full Carbon with TEST limiter for materials test
TFTR	Princeton, New Jersey, US	2.52	0.87	6.0	51		11.4	40		3	Circular Limiter	1982–1997	Full Carbon
Tore Supra	France, Cadarache	2.25	0.7	4.5(SC)	17	<2.4	9	1.7	5	2	Circular Toroidal Pumped Limiter	1988–2011	Full Carbon
WEST	France, Cadarache	2.5	0.5	3.7(SC)	17		9		7	1	D-shape Divertor	2017–	W coated C

2.2 Discharges of Current Large Tokamaks

Figure 2.1 is a photograph of the inside of TEXTOR, a limiter tokamak [1], which was built for intensive studies of PSI, together with the picture of discharged plasma in it. As seen in the figure, visible radiation from the plasma center is far less than that from boundary plasma and limiters. Because plasma temperature is high enough to fully ionize hydrogen, giving no emissions like the Balmer and the Lyman series but the Bremsstrahlung which is invisible. The Balmer series emissions from excited hydrogen atom largely contribute to radiation from boundary plasma. In addition, stronger radiation appears from the surface of the main limiter owing to their temperature rise by plasma power load.

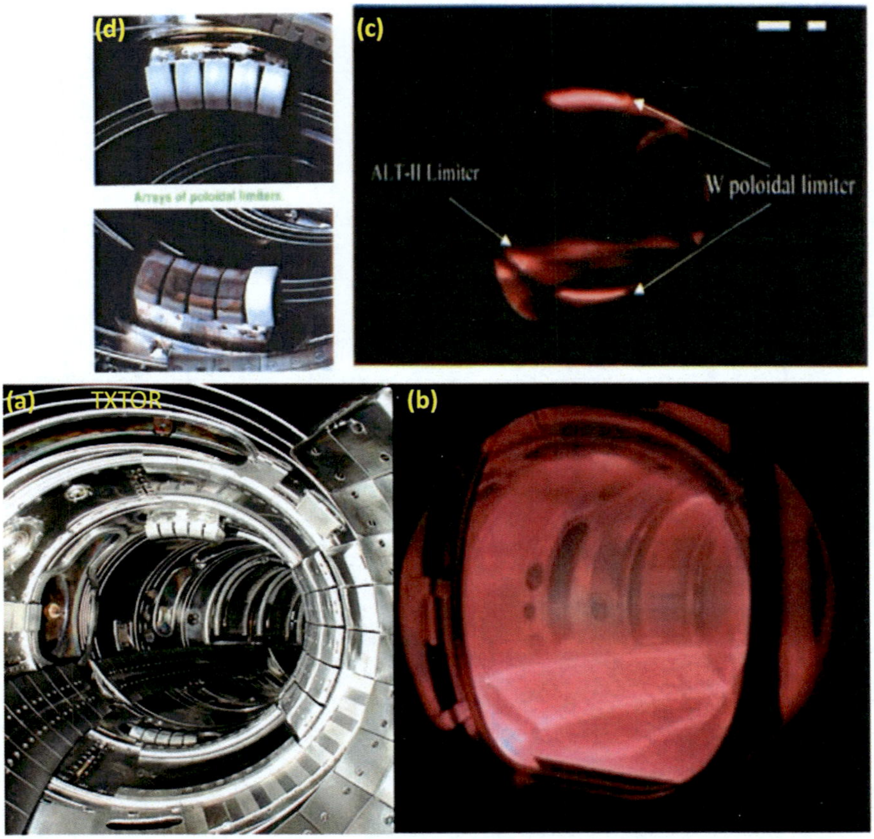

Fig. 2.1 Discharges in TEXTOR **a** inside of TEXTOR and **b** a discharge limited by ALT-II belt limiter, **c** a discharge limited with main limiters made of W-coated graphite at the bottom and top as shown in **d** with shiny one before and dark after use. During the discharge, the surface of the ALT-II belt limiter gave strong radiation owing to plasma heating

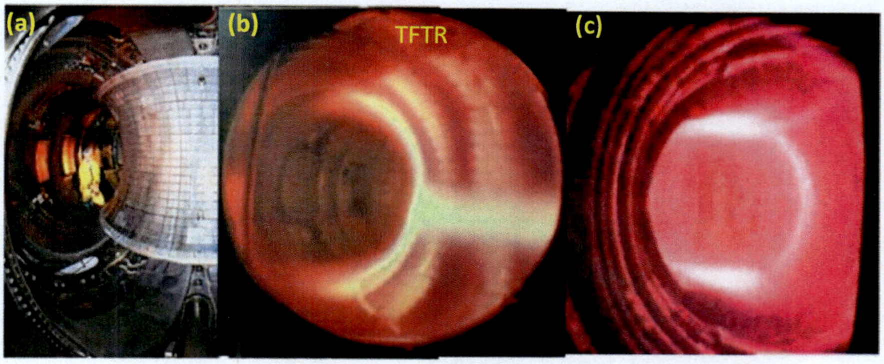

Fig. 2.2 Discharges in TFTR **a** inside of TFTR, **b** a discharge limited with inside-board (bumper limiter), and **c** a discharge limited with upper and bottom limiters

With the increasing temperature of the plasma center, the wavelength of its radiation becomes shorter from visible, UV, Soft X-rays to X-rays together with the Bremsstrahlung emission with its wavelength also shorten from IR to X-rays. Accordingly, visible radiation remains in boundary plasma and PFM. In TFTR which was also a limiter tokamak but gave higher temperature plasma, radiations from inboard surfaces working as limiters, or those from the bottom and top limiters are appreciable in Fig. 2.2 [2]. Different colors of the boundary plasma are caused by radiation from impurities, such as carbon, nitrogen, and oxygen, and the color changes depending on impurities and boundary plasma temperature. A good example of color change due to Li injection is seen in EAST as shown in Fig. 2.3 [3]. For JET (Fig. 2.4) [4], two cases are shown with emissions from both divertors and a single divertor at the bottom while the only bottom radiation in JT-60U (Fig. 2.5) [5]. In both tokamaks, radiation is mostly from the divertor area, divertor plasma, and divertor surfaces, and their color looks white indicating that the surface temperature of the divertor was quite high. Many white dots that appeared in JT-60U were caused by noises in a digital camera system due to irradiation of γ-rays and neutron produced by D-D fusion reactions.

2.3 Diagnostics for PMI Research

As shown in the previous section, optical emission is one of the most important diagnostics of plasma–material interactions (PMI). In particular, visible spectroscopy has been giving lots of information on PMI and used to be the most important diagnostic. However, with the increase of plasma temperature, emission of shorter wavelength region becomes more important and PMI is more significant. Nowadays, lots of different techniques have been employed to monitor and study PMI. Here are briefly introduced diagnostics of boundary plasmas, surfaces of PFM, and PMI.

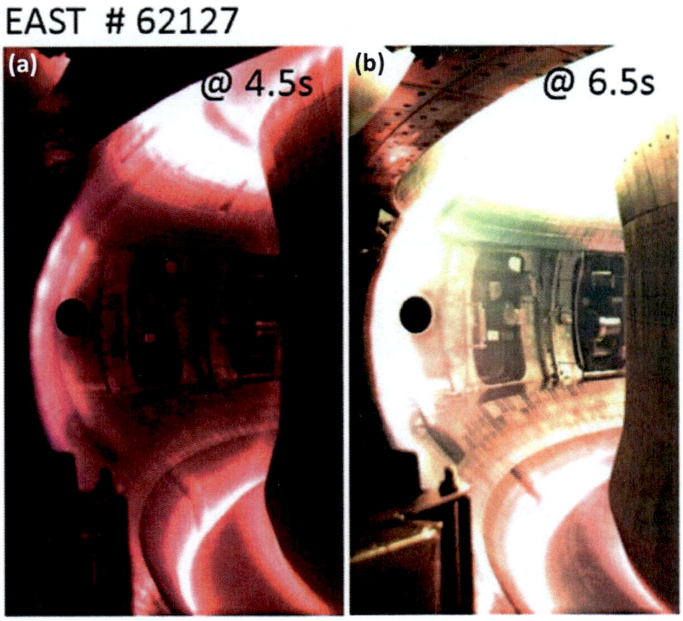

Fig. 2.3 Discharges in EAST **a** before and **b** after Li aerosol injection. The images are derived from the color-CCD camera with a field of view from window D to A of EAST (reprinted with permission from [3])

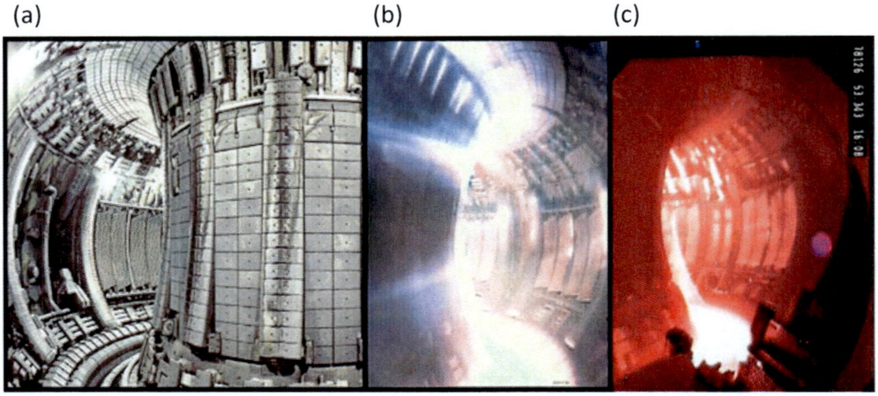

Fig. 2.4 Discharges in JET **a** inside of JET, **b** a discharge using both top and bottom divertors, and **c** a discharge using the bottom divertor. Based on images: EFDA https://fusionforenergy.europa.eu/mediacorner/imagegallery.aspx?id=24

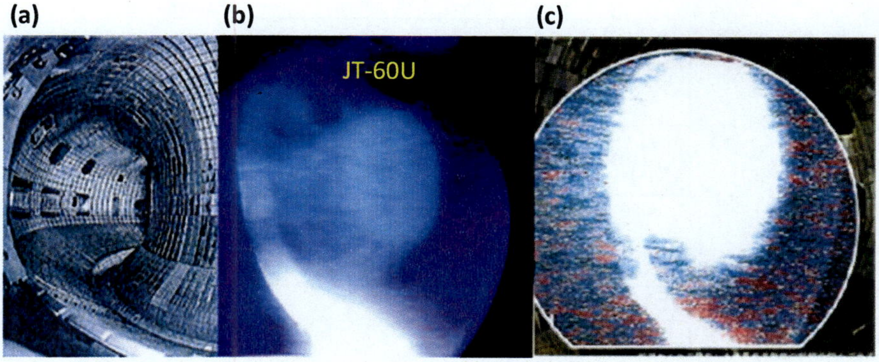

Fig. 2.5 Discharges in JT-60U **a** inside of JT-60U and **b** a discharge using the bottom W-shaped divertor. **c** A discharge significantly influenced by electric noise given by neutrons and γ-rays produced by D-D reactions

The diagnostics can be separated into two gropes, one is active observations/measurements, i.e. introducing some probes to get direct responses. The Langmuir prove is one of the most important diagnostics to measure temperature and density of electrons in the plasma, while laser irradiation for the Thomson scattering measures the electron temperature in the plasma center. Injection of high-energy ions or neutrals is an important method to determine ion temperature and density. The other is passive ones including optical measurements and collector probe measurements, i.e. let probes in targeting positions and remove them to analyze after plasma exposure, which are important methods to observe erosion and deposition of PFM and hydrogen retention.

In the following, two methods that strongly correlated to PMI, optical spectroscopy and probe measurements, are introduced.

2.3.1 Optical Spectroscopy

Optical spectroscopy is a typical passive technique, just observing radiation by a spectrometer directly or through windows or glass fibers. It is the most important diagnostic and various spectrometers are routinely used. As noted in the previous section, most of the radiations from the plasma center in current large tokamaks are in shorter wavelength than the visible region and require different techniques to detect depending on the wavelength; UV, Soft X-ray, X-ray, and γ-ray. Some examples are shown below.

Figure 2.6 shows the Balmer line emissions in the visible wavelength range from detached plasma observed in the divertor area of JT-60U [6]. From relative intensities observed in two-dimensional distributions, volume recombination was confirmed to occur in the inner divertor region of JT-60U. In boundary plasma, molecular line

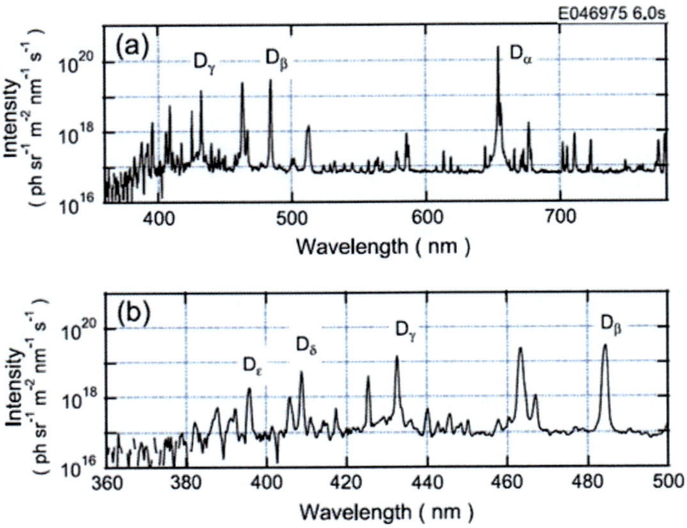

Fig. 2.6 **a** The spectrum measured in the inner divertor of JT-60U for detached plasma and **b** the expanded spectrum between 360–500 nm (reprinted with permission from [6])

emissions of hydrogen are observed as shown in Fig. 2.7 [7]. These emissions could be used for quantitative analysis of H, D, and T in boundary plasmas. However, emissions attributed to H_2, HD, HT D_2, DT, and T_2 molecules appear in narrow wavelength regions, and many emissions overlap so that qualitative analysis might not be easy.

Since, in most of the present tokamaks, central plasma temperature is high enough to fully ionize hydrogen, no emission is observed from H but from impurities. As depicted in Fig. 1–4 in Chap. 1, emission in low energy plasma with the energy of below a few hundred eV is dominated by impurities, in particular, in carbon wall tokamaks, various emissions from C impurities dominate in the VUV region as shown in Fig. 2.8 [8]. Although most of the observed emissions are well assigned to theoretically calculated ones, still the origin of some emissions is not clear because there are so many different ionized and excited states in C.

Figure 2.9 [9] shows emissions from seeded impurities of Ne and Ar in divertor of JT-60U, which are mostly in the VUV region. For heavier impurities, like W, the emission is mostly in soft X-ray to X-ray region. As seen in Fig. 1–4, for radiation from W in high-temperature plasma with the energy of 1 keV or above, radiation from W becomes significant. However, there are so many different ionized and excited states in W at higher energies, and the assignment of observed lines to particular energy states of W ions is not easy. Furthermore, it is not easy to detect spectra lines in such a short wavelength range with high-energy resolution. Figure 2.10 [10] shows observed emission from W in the plasma center of JT-60U, where estimated charge states of excited W ions are given as fractional abundance of W^{q+} ions, with q taking various numbers.

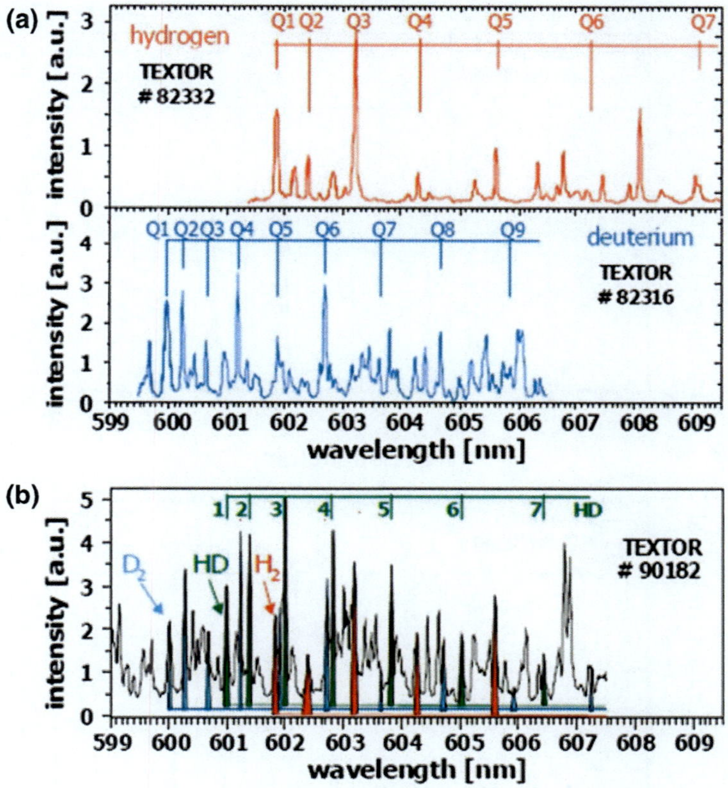

Fig. 2.7 Emission of molecular lines from TEXTOR edge plasma. Discharges of **a** pure H, **b** pure D, and **c** H+D are compared (reprinted with permission from [7])

Although loaded power by radiation to PFS can be given by the integration of intensities of the loaded radiation with wavelength, no single detector is available for that. Hence, a bolometer or calorimeter is used to detect power load to PFS, which requires some sensors or probes at locations to be observed.

It should be mentioned that although the radiation power from the plasma can be detected by the bolometer, it is not easy to find what is the main radiator (impurity). In some sense, this makes the understanding of PMI (or impurity release from PFS) difficult.

Moreover, appropriated power load test to simulate damage of PFS by radiation from burning plasma is not easy because radiation sources having equivalent energy spectra to the burning plasma which is consisted of soft X-ray to VUV are hardly available. Present power load tests are mostly done with power sources of ions, electrons, and lasers.

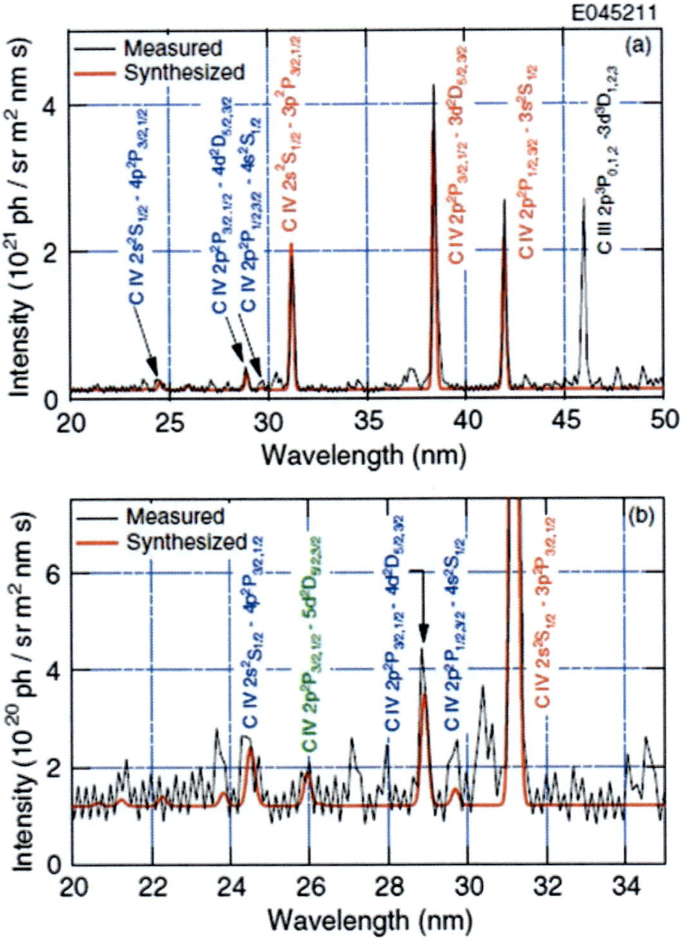

Fig. 2.8 **a** The spectrum measured with a VUV spectrometer in a wavelength range of 20–50 nm and **b** the expanded spectrum in a wavelength range of 20–35 nm for the high-X-point-configuration discharge. The thin and the thick lines indicate, respectively, the measured spectrum and the synthesized spectrum with a collisional radiative model with $T_e = 6.3$ eV, $n_e = 7.8 \times 10^{20}$ m^{-3}, and $nC^{4+}/nC^{3+} = 4.0$ (reprinted with permission from [8])

2.3.2 *Probe Measurements*

Another important method is to use probes including material test pieces, ions, electrons, laser, and so on. One of the most important is the Langmuir probe to measure electron temperature, electron density, and electric potential of a plasma. A bolometer is also a quite important probe to measure power load at the location of its sensor. Figure 2.11 shows time traces of various parameters for a TEXTOR discharge with a Mo test limiter, i.e. temperature of the limiter head, intensity of MoI line in front

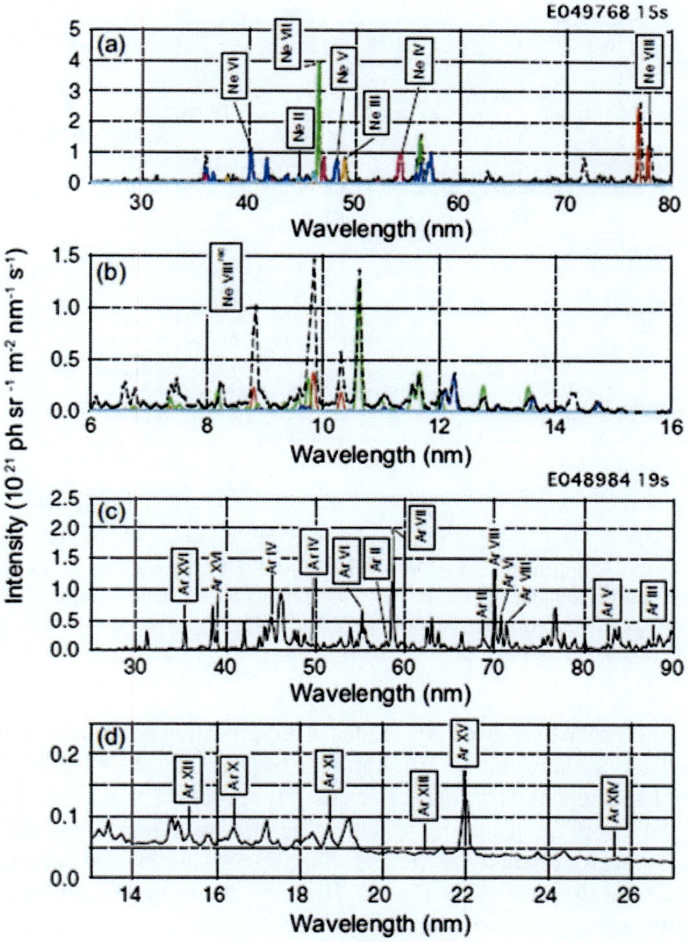

Fig. 2.9 **a** Optical spectra in UV range, and **b** spectra measured (broken line) from a Ne-seeded discharge and synthesized spectra (colored solid lines) for Ne at electron temperature of 10 eV for Ne II–IV, 20 eV for Ne V, 100 eV for Ne VI, and 200 eV for Ne VII–VIII at an electron density of 1% 10^{20} m^{-3} by Flexible Atomic Code, and **c** and **d** spectra measured from an Ar-seeded discharge. Spectral lines with a spectral symbol inside a box are used for line-radiation evaluation (reprinted with permission from [9])

of the limiter, the line average density of plasma, deuterium flux to the main limiter determined by Hα measurements, radiation measured by a bolometer, and the intensity of Mo XXVII line measured by a vacuum ultraviolet spectrometer [11]. This is the first clear indication that local impurity emission due to evaporation of Mo is separated from impurity accumulation in the plasma center. With increasing plasma temperature, power load to the probe materials is concerned. In order to avoid melting

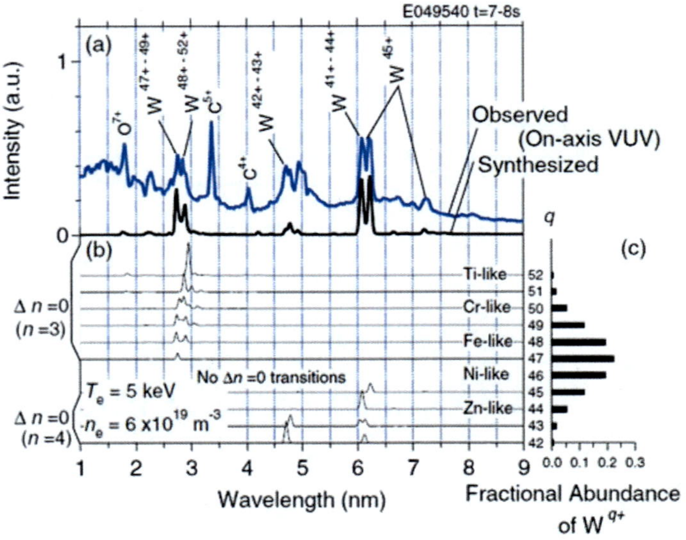

Fig. 2.10 **a** Observed optical spectra in VUV spectra in JT-60U, **b** synthesized spectra with the FAC collisional–radiative modeling for W^{q+}(q = 42–52) and **c** relative abundance of W^{q+} for the synthesis of the spectrum in (**a**). Intensity of the W^{q+}(q = 42–45) spectra in **b** is reduced to 20% (reprinted with permission from [10])

of the Langmuir probe by plasma power load, a reciprocating Langmuir probe which is introduced in plasma in only a very short time has been developed in JT-60U [12].

Beams of laser light and particles can be used for probes. For example, laser light is used for the Thomson scattering to determine electron temperature and density, and neutral beams for direct measurement of the energy of ions and neutrals. These are mainly used as diagnostics for plasma. However, they can be used for boundary plasma to give particle and radiation load on PFS.

Probes specified for PSI study are collector proves, marker tiles, and microbalance. Small test pieces are set as collector probes at specified locations in a plasma machine and exposed to plasma discharges. Afterward, they are removed and analyzed. Probes are often removable after plasma exposure or movable during plasma discharge or discharged during plasma discharge. The divertor material evaluation system (DiMES) in DIII-D is an effectively used probe for the evaluation of erosion and deposition [13]. Collector probes were extensively used in TEXTOR as summarized by Rubel et al. [14] In TEXTOR, a movable limiter was also used as a probe as mentioned above as the Mo test limiter. Depending on the purposes, the materials and structures of the probe are varied. Active use of limiters for in situ observation is employed with changing limiter materials, radial position in tokamak, and temperatures. Postmortem analysis of plasma-facing tiles or any facing components is also a powerful tool for the investigation of erosion, deposition, and hydrogen retention.

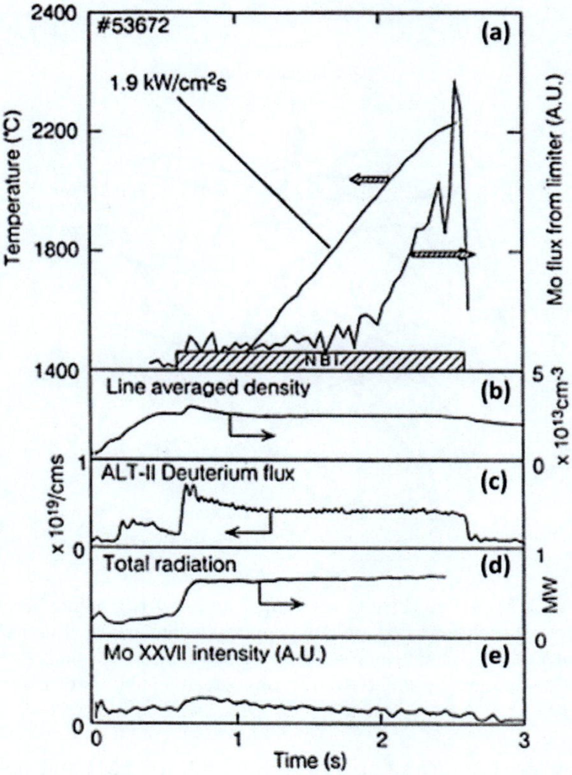

Fig. 2.11 Time evolutions of various diagnostic data obtained in a TEXTOR discharge using Mo limiter; **a** Mo flux and surface temperature near the melting surface, **b** Line averaged density, **c** deuterium flux on the main graphite limiter, **d** total radiation, and **e** Mo XXVII line intensity in the core plasma [11]

Maker tiles consisting of different materials with layered structure are often used for erosion measurements in ASDEX [15] and other devices.

2.4 PMI Observed by Proves and Limiter Experiments

PMI are different among materials used as PFM owing to differences in reflections of energy and particles, erosion, and hydrogen retention properties (dissolution, trapping, diffusion and permeation). In the early days of plasma research, except the high Z materials, like Mo and W, the effects of PFM on plasma confinement were not concerned, but low Z impurities are ubiquitous in residual gas in a vacuum. In recent large tokamak experiments using W as PFM, it becomes clear that plasma behavior is significantly influenced by PMI. In references [3–15] in Chap. 1 are described basic processes of PMI and diagnostics to observe them.

Various in situ observations of PMI have been done with using movable proves and limiter, microbalance in addition to spectroscopic measurements. In this section,

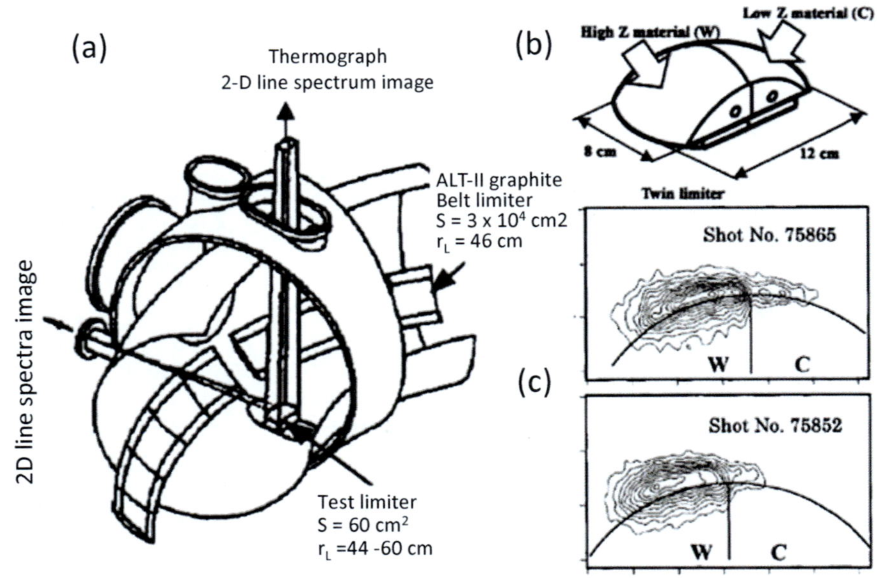

Fig. 2.12 **a** Geometry of test limiter and ALT-II graphite belt limiter in TEXTOR. The head of the test limiter was monitored by optical techniques. **b** Schematic of W/C twin limiter. **c** WI line intensity distribution in front of the W/C twin limiter for shots with the different cumulative numbers of discharges showing eroded W was spreading to C surface. [16]

some examples of in situ observation of PMI and postmortem analysis of probes intentionally set in plasma apparatus are discussed.

The effects of material difference on PMI were clearly observed in test limiter experiments carried out in TEXTOR [16–18]. Figure 2.12a is the geometry of the test limiter experiments. ALT-II belt limiter was installed in full toroidal direction at 46 cm from the plasma center and a mushroom-shaped test limiter was set at 44–60 cm and the limiter head was monitored by optical spectroscopy. Using a twin limiter which consisted of two blocks with each made of different materials, like graphite and tungsten or tantalum and tungsten as shown in Fig. 2.12b, the difference of PMI between the two different materials was examined [16]. The twin limiter was aligned in the toroidal direction facing either electron or ion drift side. For the W/C(Graphite) twin limiter, with the increasing shot number from 75862 to 75865, emission of WI lines spread to the in front of the C block indicating that eroded W is deposited on the C block in Fig. 2.12c [16]. The different reflection coefficients between C and W clearly appeared as the penetration depth of Hα into plasma as seen in Fig. 2.13 [16]. The cause of Hα emission is either electron excitation of neutral H atoms and reemitted molecules in boundary plasma or reflected H atoms at PFS in excited states. Owing to the higher reflection coefficient of W than C, the penetration length of Hα for W part of the twin W/C limiter was clearly longer as indicated in Fig. 2.13. It is also noted that the decay at a longer distance became similar indicating that the effect of the material difference in PMI mainly appears near PFS.

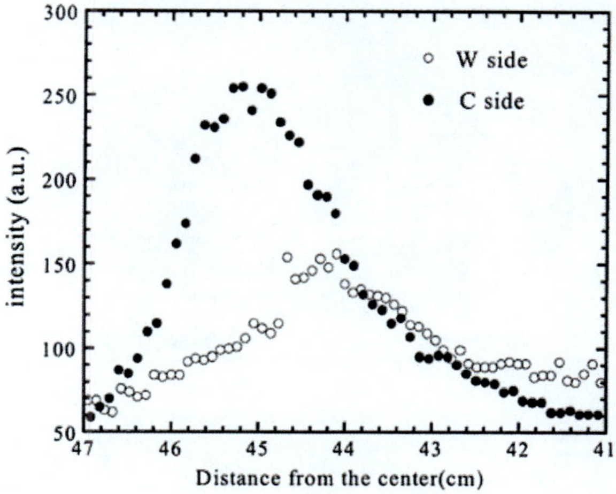

Fig. 2.13 Comparison of Dα emission intensities in front of the W/C twin limiter for the C and W sides. [16]. The limiter was set at 45.5 cm

From the picture of the C-W twin limiter after use (see Fig. 2.14a), carbon deposition at the bottom of W block was quite clear. The postmortem analysis of the limiter surfaces by RBS reveals the deposited profile of H, C, and W as shown in Fig. 2.14b [16]. Si remained as impurity after siliconization of TEXTOR [18]. As indicated by the spread of W emission in Fig. 2.12c, W deposition on C gradually decreased in the poloidal direction.

Comparison of Ta and W sides for a W/Ta twin limiter shows the difference in hydrogen retention between the two, i.e. Ta absorbs a large amount of H, while W quite small, accordingly Dγ intensity is less on the front of the Ta as seen in Fig. 2.15 [17] because the reemitted D_2 from W contributed the Dγ emission.

Postmortem analysis of probes intentionally set at various locations in plasma apparatus is one of the most important tools to observe erosion/deposition profiles and transport of eroded materials. Overview of the prove experiments in TEXTOR was given by Rubel et al. [14]. Diagnostics for studying deposition and erosion processes in JET was introduced by Coad et al. [19]. Among them, Quartz microbalance (QMB) measurements in JET [20] give a sequential change of C deposition at the location of sensors (see Fig. 2.16). Figure 2.16c [21] is an example for profiles of the deposition or erosion rates at the location of QMB indicated in Fig. 2.16a. The figure clearly shows that when the strike point was on the basal plate, erosion was high so as the deposition on QMB. In addition, H mode discharges resulted in higher erosion than L mode discharges.

DiMES in DIII-D [13] also have given suggestive information on PSI in the divertor area. Postmortem or ex-situ analysis of any materials used as components or PFS in plasma apparatus have been extensively used. Although observation by the analysis is limited to those materials exposed to the number of plasma discharges,

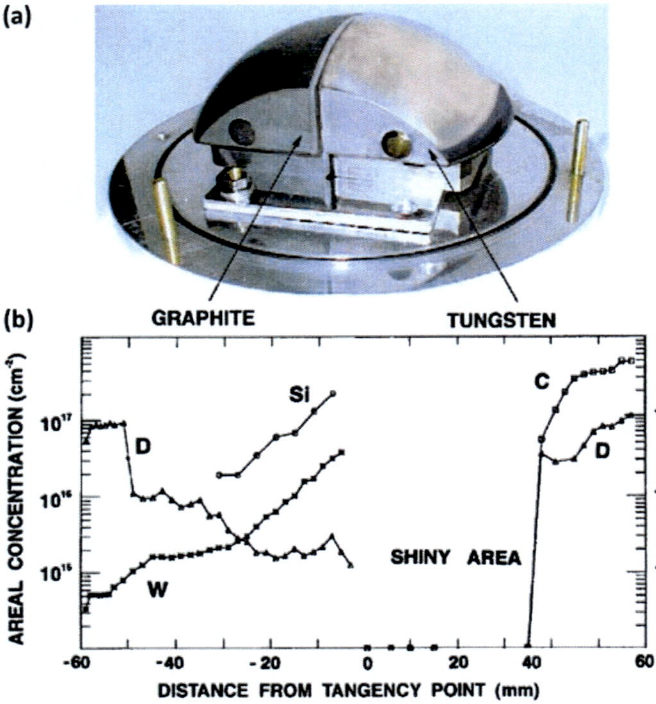

Fig. 2.14 Deposition profiles of C, W, and D on the W/C twin limiter [16]. D was mainly retained in deposited C near the bottom of both limiters. W was spreading to the C limiter, while no C deposition on shiny area of W where was erosion dominated

Fig. 2.15 Radial profiles of Dγ intensity decay in front of Ta (#86789) and W (# 86768) for Ta/W twin limiter at 1200 K. The intensities were normalized by incoming D flux onto the surface. The Dγ intensity near Ta was clearly lower than that near W [17]

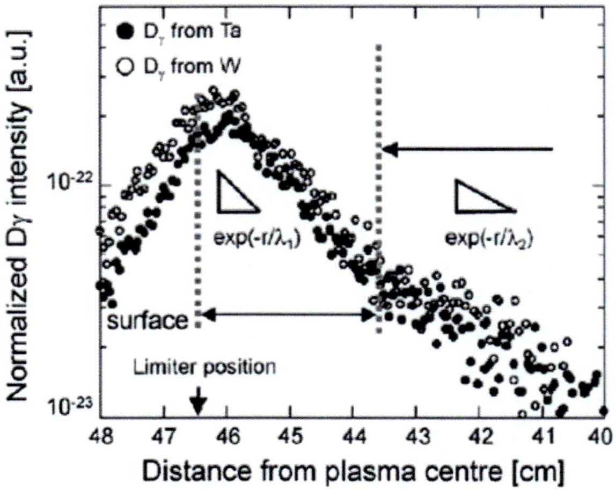

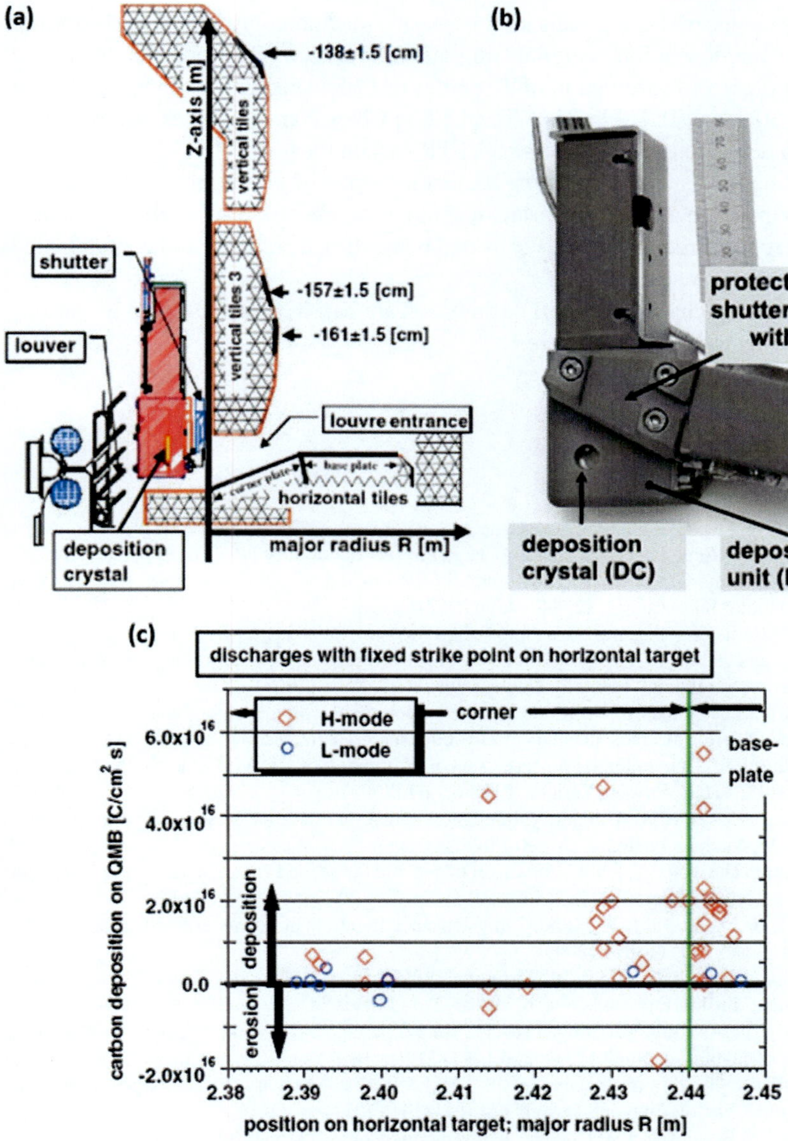

Fig. 2.16 **a** Cross section of inner divertor at the location of the QMB in JET, **b** view onto the complete system including deposition unit and shutter; shutter in the open position allows direct view onto the deposition quartz (reprinted with permission from [20]), and **c** observed deposition rates of QMB exposures with fixed strike point position on horizontal target versus strike point position (reprinted with permission from [21])

systematic choice of the materials used in plasma machines gave important information on erosion/deposition, materials migration, hydrogen retention, and so on. One good example is the observation of T profiles of PFS using tritium imaging plate in TFTR, JT-60U, and JET. Figures 5.7 and 5.8 in Chap. 5 are examples of how T was retained in deposited carbon layers on TFTR carbon tiles.

Collecting flakes and dust during the maintenance of plasma machines [22, 23] and analyzing them are also important techniques to observe erosion deposition and production of hazardous materials activated by neutron irradiation and contaminated with T.

The following chapters on PMI of this book are based on results given by various diagnostic methods described here.

References

1. Fusion Science and Technology, **47** Issue 2: Special Issue on TEXTOR (2005)
2. TFTR Public Home Page—Princeton Plasma Physics Laboratory https://w3.pppl.gov/tftr/
3. H. Mao, F. Ding, G. Luo et al., The impacts of lithium and silicon coating on the W source in EAST, Nucl. Mater. Energy **12**, 447–452 (2017)
4. JET: EUROfusion's flagship device https://www.euro-fusion.org/devices/jet/
5. H. Takenaga and the JT-60 Team, Overview of JT-60U results for the development of a steady-state advanced tokamak scenario, Nuclear Fusion **47**, S563 (2007)
6. K. Fujimoto, T. Nakano, H. Kubo, Spatial structure of volume recombination in JT-60U detached Divertor plasmas, plasma–and fusion research **4**, 025 (2009)
7. S. Brezinsek, P.T. Greenland, Ph Mertens, et al. Formation of HD Molecules in the Boundary Layer of TEXTOR, Physica Scripta, **T103** 63–67 (2003)
8. T. Nakano, H. Kubo, N. Asakura, Volume recombination of C^{4+} in detached divertor plasmas of JT-60U, Nuclear. Fusion **47**, 1458–1467 (2007)
9. T. Nakano, The JT-60 Team, Comparison of Ne and Ar seeded radiative divertor plasmas in JT-60U, J. Nucl. Mater. **463**, 555–560 (2015)
10. T. Nakano, N. Asakura, H. Kubo, Tungsten accumulation in H-mode plasmas of JT-60U, Nuclear. Fusion **49**, 115024 (2009)
11. T. Tanabe, M. Akiba, Y. Ueda, et al., On the utilization of high Z materials as plasma facing component, Fusion Eng. and Design, **39–40**, 275–285 (1998)
12. S. Sakurai, N. Asakura, N. Hosogane et al., Plasma characteristics near the X-point in W-shaped divertor of JT-60U, J. Nucl. Mater. **266**,1191–1196 (1999)
13. D.L.Rudakov, T. Abrams, R. Ding, et al., DiMES PMI research at DIII-D in support of ITER and beyond, Fusion Eng. and Design **124**, 196–201 (2017)
14. M. Rubel, S. Brezinsek, J.W. Coenen, et al., Overview of wall probes for erosion and deposition studies in the TEXTOR tokamak, Matter and Radiation at Extremes **2**, 87–104 (2017)
15. M. Mayer, V. Rohde, G. Ramos et al., Erosion of tungsten and carbon markers in the outer divertor of ASDEX-Upgrade, Physica Scripta, **T128**, 106–110 (2007)
16. T. Tanabe, T. Ohgo, M. Wada, et al., Materials mixing on W/C twin limiter in TEXTOR-94, Fusion Eng. and Design, **49–50**, 355–362 (2000)
17. T. Hirai, V. Philipps, A. Huber, et al., Hydrogen recycling and microstructures of tantalum and tungsten twin limiters exposed to TEXTOR-94, J. Nucl. Mater. **307–311**, 79–83 (2002)
18. U. Samm, P. Bogen, G. Esser et al., Plasma edge physics with siliconization in TEXTOR, J. Nucl. Mater., **220–222**, 25–35 (1995)
19. J.P. Coad, H.-G. Esser, J. Likonen, et al., Diagnostics for studying deposition and erosion processes in JET, Fusion Eng. Design, **74**, 745–749 (2005)

20. H.G. Esser, G. Neill, P. Coade, et al., Quartz microbalance: a time resolved diagnostic to measure material deposition in JET, Fusion Eng. and Design, **66–68**, 855–860 (2003)
21. H.G. Esser, V. Philipps, M. Freisinger et al., Effect of plasma configuration on carbon migration measured in the inner divertor of JET using quartz microbalance, J. Nucl. Mater. **337–339**, 84–88 (2005)
22. J.P. Sharpe, D.A. Petti, H.-W. Bartels, Review of dust in fusion devices: Implications for safety and operational performance. Fusion Eng. and Design **63**(64), 153–163 (2002)
23. A. Baron-Wiechec, E. Fortuna-Zaleśna, J. Grzonka, et al., First dust study in JET with the ITER-like wall: sampling, analysis and classification, Nuclear Fusion, **55**, 113033 (7 pp) (2015)

Chapter 3
Power Load on Plasma-Facing Materials

3.1 Introduction

In the early days of research on plasma–material interactions (PMI), the main research targets were the spontaneous response of material surface or surface phenomena induced by the collision of incident energetic particles, such as reflection, secondary particles emission, photon emission, and reemission of incident particles, which are discussed in Chap. 4. In the progress of plasma confinement with higher temperature and density, power load to plasma-facing materials (PFM) becomes so high to cause materials melting, erosion, and degradation of materials properties. Consequently, the lifetime of PFM becomes concern. Influences of power load to the materials appear as sublimation, melting, or re-solidification which sometimes accompanies material cracking or flaking, and heat-influenced zone in near surface such as recrystallization. In reverse, melting and evaporation influence plasma confinement by plasma contamination and melt layer motion. Ironically, extremely high-power load accompanying a significant amount of material sublimation could protect further power load by vapor shielding which is a new topic.

This chapter is devoted to describe how power is loaded to PFM and what carries the power. In Sect. 3.2, the distribution of power load to various locations in a fusion reactor is briefly introduced. Since the power load is divided into two categories, i.e. steady state and transient or off-normal event like disruption, they are separately introduced in Sects. 3.3 and 3.4, respectively. Power load by a neutron is also separately discussed in Sect. 3.5. The detailed material responses to the power loads are described in Chap. 5.

T. Tanabe, *Plasma-Material Interactions in a Controlled Fusion Reactor*, Springer Series in Plasma Science and Technology, https://doi.org/10.1007/978-981-16-0328-0_3

3.2 Estimation of Power Load and Its Distribution in a Fusion Reactor

A very simple calculation gives power load to plasma-facing surface (PFS) in a fusion reactor with a thermal output power of 3 GWth or electrical output of 1 GWe in Table 3.1. In the table, the reactor is assumed to be spherical for simplicity with its radius of 3 m or 5 m of which volume and surface area of the reactor are 113 m^2 and 113 m^3, and 314 m^2 and 523 m^3, respectively. Accordingly, averaged power flux and produced energy density are 26.5 MW m^{-2} and 26.5 MW m^{-2}, and 5.7 MW m^{-2} and 9.5 MW m^{-2}, respectively. Nearly 2/3 of the output power is carried by neutron and is deposited in a blanket, while 1/3 (8.8 MW m^{-2} or 3.2 MW m^{-2}) is carried by energetic ions and neutral, electrons, and photons and deposited to PFS. Since any materials have their maximum tolerable power load to avoid melting damage, around 10–20 MW m^{-2} under intensive cooling, the power load is one of the limiting factors on designing the reactor size. The larger is better to reduce the power load, while the smaller is better for energy density increasing energy efficiency. This limits the minimum size of a reactor. For example, it is unrealistic to build a power reactor with 3 GWth having a diameter of less than 2.5 m.

In Fig. 3.1, power loads to PFS and blanket are specified for a tokamak reactor having a divertor structure with the output power of 3 GWth under steady-state operation. Nearly 80% of the fusion power, 2.4 GW, is carried by 14 MeV neutrons and deposited in the blanket to be converted to heat or electric power and to breed tritium. The remaining 20%, 600 MW, is carried by energetic ions and neutral, electrons, and photons. In addition, to sustain the burning plasma, external heating of around 50 MW is put in. Hence, the power of 650 MW should be exhausted as radiations and particle energies to PFS. Around 500 MW is carried by He ions produced by D-T reaction and used to sustain high temperature burning plasma, which turns to the core radiation mostly by Bremsstrahlung and some by impurity radiation as seen in Fig. 1.4. The core radiation is rather homogeneously loaded to PFS resulting in the power load of 0.5 MW m^{-2} on average. The remaining 150 MW are exhausted as radiation and particle energies, a large part of which are loaded to the divertor region. Consequently, the power loads to the first wall and the divertor

Table 3.1 Simple estimation of power load and energy density for spherical reactors with thermal output of 3 GWth with radius of 3 or 5 m

Radius (r) m	Surface area ($4\pi r^2$) m^2	Power flux MW m^{-2}	Volume ($4\pi r^3/3$) m^3	Energy density MW m^{-3}
3	113	26.5	113	26.5
5	314	9.5	523	5.7

2/3 of output power (6-16 MW m^{-2}) is carried by 14 MeV neutrons and deposited in a blanket having large volume, and 1/3 by particles and radiation (3–10 MW m^{-2}) and deposited to plasma-facing surface

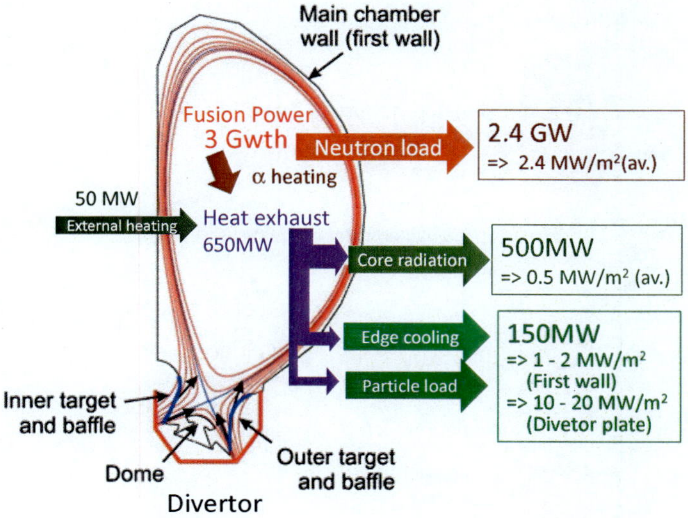

Fig. 3.1 Estimation of power load or power exhaust to PFS in a fusion reactor with an output power of 3 GWth under steady-state operation

become 1–2 MW m^{-2} and 10–20 MW m^{-2}, respectively. A more detailed estimation of power load was made for ITER and DEMO in references [1, 2], respectively, for examples.

Figure 3.2 compares power load or emission in various energy systems. Steady-state power load and transient power load are separated. In a fission reactor, the main part of the energy released by fission reactions in fuel pins is removed from their surfaces by cooling water surrounding them. Because of the material limitation or to avoid the melting or the degradation of materials parameters of the fuel pins, the power flux from the fuel pins is controlled to be below a few MW m^{-2}. The power flux of a boiler of a normal power plant is one order of magnitude less. Power load to divertor surfaces in a fusion reactor should be designed to be below the melting/evaporation threshold. To avoid melting damage, various efforts for cooling have been done. Figure 3.3 is an example of power load tests (Fig. 3.3b) using ITER divertor mockup with W hot pressed on a Cu alloy heat sink with a swirl tube for efficient cooling with different joining conditions (Fig. 3.3a). In the best condition, the surface temperature of the mockup under a power load of 13 MW m^{-2} was kept below 1500 K, the recrystallization temperature of W avoiding any thermal damage [3].

In addition to the steady-state power load, some transient power loads given by instability of plasma, plasma collapse, or disruption releasing most of the energy confined in the burning plasma within a very short time are not likely avoided. The effect of the transient power load on materials depends critically on its time duration and loaded area. Protective layers of a spaceplane are subjected to large heat load caused by friction and integrated power load with time is quite high (see Fig. 3.2). On the other hand, the time duration of the transient power load in a fusion reactor

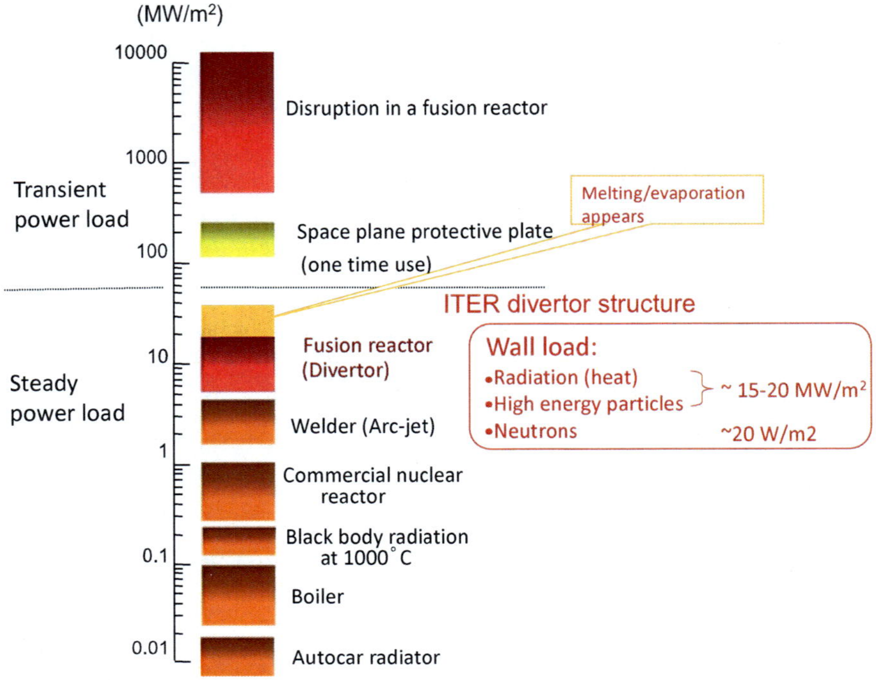

Fig. 3.2 Comparison of surface power load in various energy systems

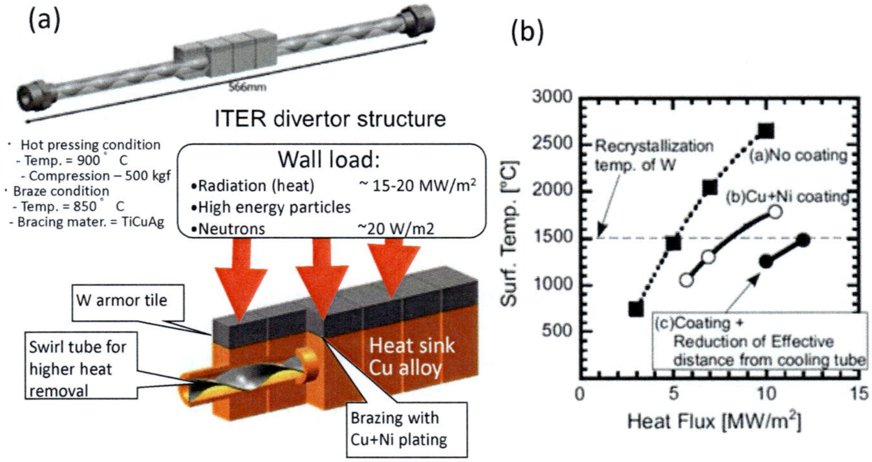

Fig. 3.3 Power load test for ITER divertor mockup with W hot pressed on a Cu alloy as a heat sink with a swirl tube for cooling with different joining conditions. **a** Geometrical structure of the mockup and **b** comparison of thermal response of W hot-pressing mockups with different joining conditions in high-power flux experiments (reprinted with permission from [3])

is much shorter than that given on the spaceplane. Accordingly, the transient power load in a fusion reactor becomes much higher and easily causes ablation or material damages.

Materials' tolerance to such high-power load depends on their thermo-physical properties, such as thermal conductivity, melting point, heat capacity, thermal shock resistance, and so on as described in Chap. 4. In addition, these material parameters are degraded by power loads especially neutron load.

Candidate PFM having the highest melting point can tolerate the power load of about 20 MW m^{-2} under intensive cooling but not more. Therefore, two critical issues are remaining, one is the reduction of the power load to PFS to be less than their tolerant level by cooling boundary plasma, and the other is the development or selection of PFM. At present, even for the highest melting temperature materials, W and C being used as armor tiles, the power load to the armor tiles in a fusion reactor seems high enough to damage them. Plasma detachment with high divertor pressure and impurity seeding [4] to enhance radiation cooling is encouraged. Still, they are not enough to ensure the robustness of the divertor.

Power load by particle fluxes are another concern. Suppose 1/3 of the power flux of 1 MW/m^2 was carried by fuel particles with an energy of around 100 eV, fuel particle fluxes would be 6×10^{22} m^2 s^{-1}. Based on the power load given in Fig. 3.1, the fuel particle flux to the first wall and divertor plate would be 10^{23} and 10^{24} m^2 s^{-1}, respectively. Compared to the surface atomic density of a solid material (around 10^{19} m^{-2}), at least every 10^{-4} s to 10^{-5} s, one plasma fuel particle passes nearby or collides with a surface atom. If the surface atom gets energy from the plasma particles by collision, it takes a millisecond or more to be relaxed. This means that all surface atoms are not in thermal equilibrium with bulk temperature but at excited or higher temperature states. Furthermore, such excited atoms or higher temperature atoms could emit photons (radiation) to be relaxed. Therefore, the mitigation of the power load to the divertor plates by particles is also mandatory. Power exhaust from burning plasma and mitigation of power load to PFS are the most important targets in recent PMI researches.

3.3 Steady-State Power Load

At steady state or normal operation, various particles carry power to PFS: charged particles including fuel ions, He and other impurity ions, energetic neutrals produced by charge exchange, electrons, neutrons produced by D-T reactions, and radiation (photons) from plasma. In addition to the steady power load, periodic power load caused by saw tooth, blobs, and edge localized mode (ELM) activities are superposed. These power loads are often very much localized to a specified area, and sometimes resulting in extremely high-power load in the narrow area. In particular, very large ELM (often referred to as giant ELM) is concerned [5]. The giant ELM is discussed as a transient power load in the next section.

As noted in Fig. 1.4, the energy of radiation distributes vary widely from characteristic X-rays of impurities in plasma, bremsstrahlung from burning plasma which is also widely distributed from X-ray to infrared, UV-Viz and infrared range photons from boundary plasma, and visible radiation from high-temperature material surfaces.

Material response to the radiation changes with its photon energy. Some uncertainty remains in the material response to high-energy photons with high flux. Powerful photon sources are hardly available to simulate the power load in a fusion reactor. In particular, photon sources with energy ranging from soft X-ray to UV are scarce. The photons with such an energy range interact strongly with any elements, resulting in secondary phenomena, emissions of secondary ions, electrons, and photons. Consequently, PMI phenomena in plasma boundary must be significantly modified by such high-energy photon irradiation. Particularly opacity change in the boundary plasma caused by the secondary phenomena would influence plasma confinement and power load to PFS. Therefore, in some sense, ITER divertor is a good testbed for the investigation of PSI in a fusion reactor.

Power loads by ions, electrons, and charge exchanged neutrals are controlled by electric–magnetic field or plasma configuration, and hence depending on the location of PFS, PSI phenomena appear differently. In normal operation, the highest power is loaded at the divertor target position and nearby the outer divertor target, and the power load to the inner first wall surface is higher than that of the outer first wall surface.

Although some very high energy of T and He produced by D-T reactions are escaping from burning plasma owing to ripple loss or orbital loss, their power would not be large enough to influence the total power load to PFS. Nevertheless, they could induce radiation damage in near surface regions of PFM and influence fuel recycling as discussed in Chap. 7.

In addition to the above described steady power loads, periodically changing power load is added, they are given by saw tooth activity, blobs, and ELM. Although the power load given by saw tooth activity is not likely very large, ELM activity is found to be sometimes very large in high-density and high-temperature plasma. Now giant ELM concerns seriously [6, 7], which appears rather irregularly and discussed in the next section.

3.4 Transient Power Load

Typical transient power load is given by ramp-up and ramp-down phase of plasma, particularly in the later, energy used for confinement is dissipated to PFM, and control of slower rump-down is quite important. It is well known that during the ramp-up phase, lots of impurities (mainly water) adsorbed on PFS are released because some of the input power for plasma heating is dissipated to PFS to heat up its surface temperature. Sometimes when wall conditioning is not well done, plasma collapses during the ramp-up phase.

The sudden collapse of plasma confinement results in a disruption in which most of the stored energy in plasma is loaded to a localized area of PFS in a very short time, hence, resulting in extremely high-power load. The disruption often accompanies the emission of runaway electron beams with an energy of MeV range within a very short time and could destroy plasma-facing components and reactor vessel. If the energy of 100 MW is given to the area of 1 m^2 within 1 ms, its loaded power to the area becomes 100 GJ m^{-2}, which exceeds the melting threshold of any materials. Mitigation of disruption is one of the most important tasks to establish a fusion reactor as an energy source [8, 9]. Another concern is ELM which seems necessary to keep plasma confinement resulting in pulsed power load to the localized area of PFS. Sometimes the power load by ELM exceeds 1 MJ m^{-2} as observed in giant ELM in JET and ELM ablation would limit ITER divertor lifetime as indicated in Fig. 3.4 [10]. Compared to CFC (Carbon Fiber enforced Carbon) of which material loss is mainly caused by radiation-induced sublimation and chemical sputtering, W seems to allow higher ELM energy density if it is not melted. However, materials' loss of W by melting significantly reduces acceptable numbers large of ELMs. Hence mitigation of ELM power load is quite important to avoid melting of W. Materials' responses to high-power load are discussed separately in Chap. 5.

The localized power load influences hydrogen behavior because of temperature rise. However, the effects of such high-pulsed power load on hydrogen recycling have not been investigated well. In current tokamak, plasma confinement relies on a low hydrogen recycling mode, i.e. most of fuel particles incident to PFS is retained, and

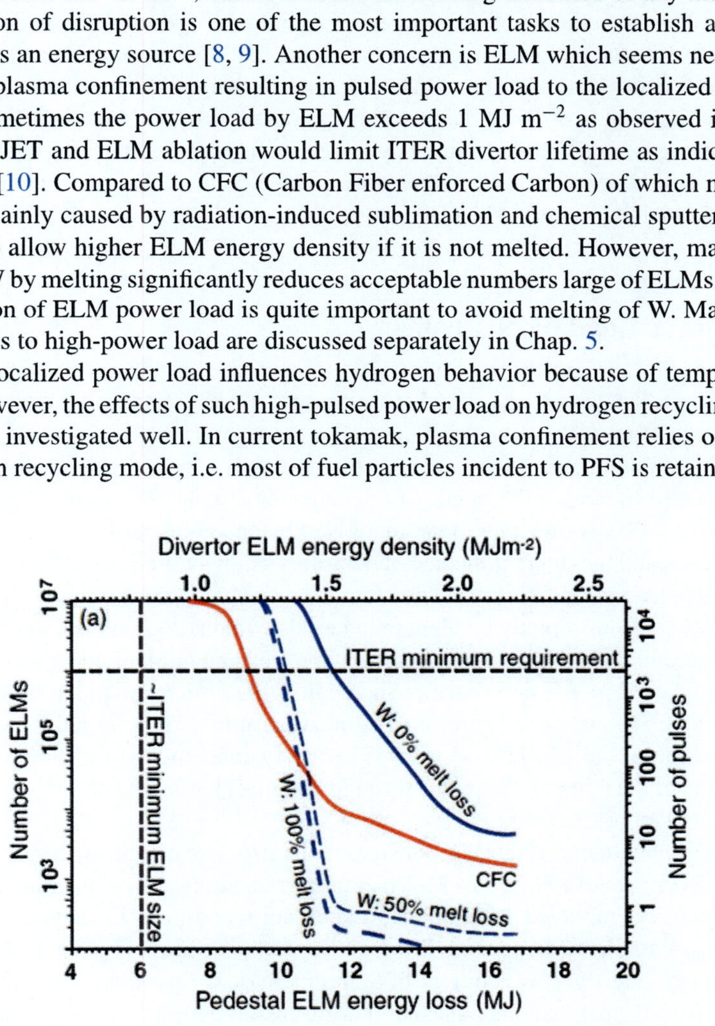

Fig. 3.4 Estimated erosion by ELM power load: **a** predicted CFC (20 mm starting thickness) or W (10 mm thickness) divertor target lifetime due to ablation expressed in terms of the number of ELMs or number of full power ITER pulses ($f_{ELM} = 1$ Hz) as a function of pedestal energy loss per ELM or divertor energy density for an inter-ELM power density of 5 MW m^{-2} including the case of 0, 50%, or 100% melt layer loss for W (reprinted with permission from [10]). See references therein

wall saturation, i.e. no retention of the incident particles makes plasma density control difficult. However, the wall saturation is unavoidable after a long discharge duration. Temperature increase during the discharge also makes the wall saturation earlier owing to the reduction of the saturation concentration with increasing temperature, which makes plasma density control difficult.

In current tokamak experiments, their discharge duration has been limited within a few minutes, i.e. steady power and particle balance have not been established well. In most tokamaks, cooling of vacuum vessel is ambient. Therefore, some tokamaks require breaks between discharges of more than a half hour to cool down the vessel or magnetic coils. In a fusion reactor, continuous discharge is planned, even in ITER discharge duration is 400 s. Gradual temperature increase could cause sudden fuel release owing to sudden detrapping of hydrogen such like decomposition of chemical compound. The effects of wall temperature change on boundary plasma are one of the main remaining issues to be clarified.

3.5 Power Load by Neutrons

As indicated in Fig. 3.1, the power load by D-T fusion neutrons, 2.4 MW m^{-2}, is not small. However, the neutrons penetrate through plasma-facing wall into the blanket region and deposit their energy volumetrically. Then, the deposited energy is transformed to heat in the blanket to generate electricity. Therefore, the neutron power load to PFS is much less than that given by energetic particles and radiation from plasma, and the direct influence of neutron loading on PFS and PMI would not be significant.

14 MeV neutrons directly incident to materials without deceleration in water like a fission reactor and give both displacement and transmutation of constituent atoms in materials used in a fusion reactor. Such damaging effects are much higher than those in the fission reactor. Therefore, degradation of material properties caused by the irradiation of the 14 MeV neutrons is seriously concerned [11]. They will limit the lifetime of the structure materials through materials hardening or loss of ductility and resultant embrittlement.

Comparing with the volumetric damages in the structure materials, degradation of material properties of PFM by 14 MeV neutron irradiation is less concerned because they are usually employed as thin plates and do not function as a structure material. However, accumulated damages in PFM work as trapping sites of hydrogen fuels and consequently, alter fuel recycling at PFS and increase T retention. Since this book focuses on PMI, materials damages are not discussed except those appearing at or near PFS to influence PSI.

It should be noted that T breeding and recovery in the blanket system shall be significantly influenced by additional damages accompanied by nuclear reactions for T breeding, i.e. the energy of 14 MeV neutron is converted to heat and used to breed tritium through the reactions with ^{6}Li and ^{7}Li,

$$^{6}\mathrm{Li} + \mathrm{n} \rightarrow \mathrm{T} + {}^{4}\mathrm{He} + 4.8\mathrm{MeV}$$

$$^{7}\mathrm{Li} + \mathrm{n} \rightarrow \mathrm{T} + {}^{4}\mathrm{He} + \mathrm{n} - 2.5\mathrm{MeV}$$

The reaction products of T and ^{4}He have enough energy to displace constituent atoms of the breeding materials and damages produced in the breeding materials trap T and make thermal recovery of T harder. At the same time, γ radiation given by the reaction (2.1) and induced by neutron activation of structure materials increases the temperature to enhance the chemical process in the blanket system.

Although T breeding and recovery in the blanket to fulfill the fuel self-sufficiency are key issues for the establishments of a fusion reactor as an energy source, they are not discussed here. One can refer a book concentrating T as a fusion fuel [12].

3.6 Mitigation of Power Load (Power Exhaust)

As discussed in Sect. 3.2, the power used to sustain burning plasma should be exhausted as radiation and particle energy and loaded to PFM. However, the power load should be kept below the threshold of material damage. Hence, the amount of loaded power, areal size, and time duration are limited. In steady-state operation, the power load to the divertor area is the highest. The power load reduction has been tried with either or both of the reduction of the particle flux and dissipation of radiation to a wider area. To realize the former, increasing the divertor target area is straightforward and making W-shaped divertor to have longer legs is a suitable way, although it makes the divertor structure complex. To enhance the radiation in front of PFM, impurity seeding is studied [4], in which an inert gas with medium atomic (Z) number, such as N_2, Ne, and Ar, is introduced in boundary plasma as an impurity.

High radiation from impurites used to be troublesome to reduce plasma temperature when plasma temperature was under 1 keV as depicted in Fig. 1.4. For higher temperature plasma over 5 keV, central radiation from low Z impurities is not significant, while their radiation in boundary plasma becomes significant. In C-wall tokamaks, significant radiation from eroded C was observed as Carbon blooms in JET and TFTR [13] and Carbon MARFE in JT-60U [14], of which details are given in Sect. 4.7 in Chap. 4. This is a typical example showing that research targets or physics and chemistry in plasma change with plasma temperature or with the development of plasma confinement, as discussed in Chap. 1.

Formation of detached plasma [15] in divertor area is another way, which is realized with an increase of gas pressure under appropriated control of pumping speed and additional gas puffing either fuel gas or impurity gases.

The transient power load should be kept low enough to avoid the destruction of plasma-facing components including material damages of PFM. Since disruption would give the most severe damages, lots of effort have been done to avoid the disruption and to mitigate the material damage [16, 17]. Still, disruption-free plasma is not likely established in ITER and should be one of the most important research subjects in ITER. Recently, effects of runaway electrons, which are always accompanied with the disruption and appear at many plasma transient events, are concerned and their suppression is tried [18]. Since its impact on PFM of present tokamaks has not been serious. So how to simulate the effects of runaway electrons on PFM also remain as a research subject.

References

1. R.A. Pitts, S. Carpentier, F. Escourbiac et al., Physics basis and design of the ITER plasma-facing components. J. Nucl. Mater. **415**, S957–S964 (2011)
2. R. Wenninger, R. Albanese, R. Ambrosino et al., The DEMO wall load challenge. Nuclear Fusion **57**, 046002 (11 pp) (2017)
3. K. Ezato, S. Suzuki, K. Sato, M. Akiba, Development of DEMO Divertor with Reduced Activation Ferritic/Martensitic Steel (F82H) in JAEA, Book of Abstracts, FT/P5-37, 226 p. IAEA fusion energy conference; Chengdu (China); 16–21 Oct 2006
4. A. Kallenbach, M. Balden, R. Dux et al., Plasma surface interactions in impurity seeded plasmas. J. Nucl. Mater. **415**, S19–S26 (2011)
5. G. Federici, A. Zhitlukhin, N. Arkhipov et al., Effects of ELMs and disruptions on ITER divertor armour materials. J. Nucl. Mater. **337–339**, 684–690 (2005)
6. Yu.F. Baranov, C.D. Challis, J. Ongena, et al. Large ELM-like events triggered by core MHD in JET advanced tokamak plasmas: impact on plasmas profiles, plasma-facing components and heating systems. Nuclear Fusion **52**, 023018 (14 pp) (2012)
7. R.A. Pitts, G. Arnoux, M. Beurskens, The impact of large ELMs on JET. J. Nucl. Mater. **390–391**, 755–759 (2009)
8. E.M. Hollmann, P.B. Aleynikov, T. Fülöp et al., Status of research toward the ITER disruption mitigation system. Physics Plasmas **22**, 021802 (2015). https://doi.org/10.1063/1.4901251
9. L.R. Baylor, C.C. Barbier, J.R. Carmichael et al., Disruption mitigation system developments and design for ITER. Fusion Sci. Technol. **68**, 211–215 (2015)
10. R.A. Pitts, J.P. Coad, D.P. Coster, Material erosion and migration in tokamaks. Plasma Phys. Control. Fusion **47**, B303–B322 (2005)
11. V. Barabash, G. Federici, J. Linke et al., Material/plasma surface interaction issues following neutron damage. J. Nucl. Mater. **313–316**, 42–51 (2003)
12. T. Tanabe, *Tritium: Fuel of Fusion Reactors*. Springer Japan (2017). https://doi.org/10.1007/978-4-431-56460-7
13. M. Ulrickson, The JET Team, The TFTR Team, A review of carbon blooms on JET and TFTR. J. Nucl. Mater. **176 & 177**, 44–50 (1990)
14. T. Nakano, H. Kubo, N. Asakura et al., Radiation process of carbon ions in JT-60U detached divertor plasmas. J. Nucl. Mater. **390–391**, 255–258 (2009)
15. A.W. Leonard, Plasma detachment in divertor tokamaks. Plasma Phys. Controlled Fusion **60**, 044001 (2018)
16. M. Lehnen, A. Alonso, G. Arnoux et al., Disruption mitigation by massive gas injection in JET. Nucl. Fusion **51**, 123010 (2011)
17. E.M. Hollmann, P.B. Aleynikov, T. Fülöp et al., Status of research toward the ITER disruption mitigation system. Physics Plasmas **22**(2015)021802. https://doi.org/10.1063/1.4901251
18. R.S. Granetz, B. Esposito, J.H. Kim et al., An ITPA joint experiment to study runaway electron generation and suppression. Phys. Plasmas **21**, 072506 (2014)

Part II
Basic Processes in PMI

Chapter 4
Responses of Plasma-Facing Surface to Power Load Given by Radiation and Energetic Particles

4.1 Introduction

Material surfaces exposed or facing to plasma are subjected to radiation (energetic photons) and particles including ions, neutrals and electrons, and residual gases in a reactor vessel. Injected photons and particles except directly reflected ones lose their energy by collisional processes in materials until they fully lose their energy to stop at a certain depth which changes depending on the incident energy. As the consequence of such energetic particle injection, secondary particles and photons, in addition to their reflection/scattering, are emitted from the surface and near surface of the materials, and the materials are modified with injected particles remained and various defects caused by collisional processes of the incident particles and their constituent atoms.

Figure 4.1 schematically shows basic processes of ion solid interactions [1]. Although the driving force of electromagnetic interaction of electrons with materials is generally the same as that of the ions, their lighter mass results in less significant effects. Except for quite high-energy injection (>~ MeV), electrons do not displace constituent atoms of the materials. Therefore, the influence of electron impact on plasma-facing materials (PFM) is far less than that given by ion impact. However, in a fusion reactor, the electron impact would have a certain role as emissions of secondary electrons and photons.

It should be noted that depending on transferred energy in each phenomenon in Fig. 4.1, its time duration changes, from initial ion incidence within picosecond to nanosecond to thermalization within second to hour as shown in Fig. 4.2 (See also Fig. 1.3). The figures also show characteristic times and transferred energies in physical and chemical events on atomic scale. Details of the collisional process of constituent elements of materials with injected energetic ions including neutrons are described in various textbooks relating radiation damages like, for example, references [2–4]. This chapter deals mainly with these phenomena relating to plasma–material interactions (PMI), i.e. occurring on/in PFM and other components in a reactor vessel.

T. Tanabe, *Plasma-Material Interactions in a Controlled Fusion Reactor*, Springer Series in Plasma Science and Technology, https://doi.org/10.1007/978-981-16-0328-0_4

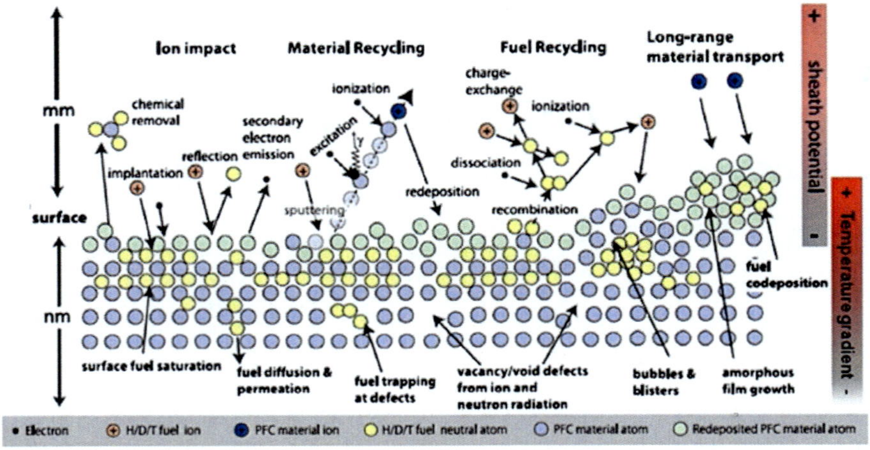

Fig. 4.1 Schematic illustration of the synergistic plasma–surface interaction processes that dictate material evolution and performance in the magnetic fusion plasma environment (reprinted with permission from [1])

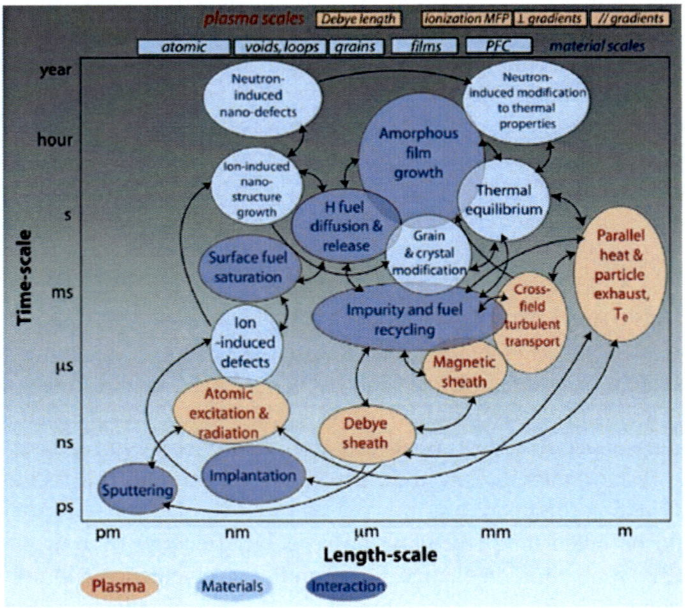

Fig. 4.2 Graphical representation of the multiple time and length-scales involved in the inherently coupled processes and phenomena that dictate plasma–materials interactions in the boundary plasma region of magnetic fusion devices. Processes occurring within the plasma are denoted in light red, while those in the near surface and bulk materials are in light blue, and the important plasma materials interactions are identified in light purple (reprinted with permission from [1])

Various particles including charged and noncharged particles, electrons, photons, and even phonons are emitted from the materials' surface subjected to energetic photons and particles. In this chapter, following the brief introduction of the energy loss process of injected particles in Sect. 4.2, four subjects are separately described, which are emissions of ions and neutrals in Sect. 4.3, emission of electrons and photons in Sect. 4.4, energy reflection in Sect. 4.5, and reemission of incident ions in Sect. 4.6. Although the emission of phonons is always accompanied by energetic particle injection, their effects on PMI are quite small and not discussed here. However, it should be mentioned that detection of the phonons could be used as one of the diagnostic methods to investigate boundary plasma and hence encouraged to make more detailed investigation in future. In Sect. 4.7, interactions of released particles from PFM with photons and electrons in boundary plasma are briefly explained as bases of PMI [5].

In PMI studies, the main subject is to understand boundary plasmas and the influence of plasma-facing surfaces (PFS) on them. Simulations of boundary plasma, and erosion and deposition are well developed as various codes like B2-EIRENE [6], ERO [7], PARASOL [8], Integrated SOL/Divertor Code [9], SOLPS-ITER [10], and UEDGE [11]. Some of them are routinely used in designing ITER and reactors.

In some codes, erosion and deposition and recycled hydrogen are included. However, some phenomena of PMI discussed in this book are not included, and the behavior of recycled or reemitted hydrogen is very much simplified. Recycled hydrogen includes various species, molecule, atoms, their excited states, and ions with a variously different energy. Therefore, for edge simulation, fluxes and energy distributions of these species should be included. One of the aims of this book is to demonstrate that PMI phenomena considered form material sides are more important than that have been considered up to now.

4.2 Energy Loss Processes of Energetic Particles Injected in a Solid Target

Figure 4.3 schematically shows energy loss processes of energetic particle injected in a solid in micro-scale, which are consisted of two processes. If an incident particle is neutral and its energy is high enough to ionize itself, it is immediately ionized at the surface. Therefore, differences in the energy loss process between neutral and ions injections are quite small. Hereafter incident particles are assumed to be ions. Initially, electron excitation or ionization of target atoms dominates energy loss. The energy loss of an incident ion with energy E is inversely proportional to penetration depth (dE/dx), which is nearly constant. Because of this constant energy loss region, the depth dominated by the ionization is referred to as the linear energy transfer (LET) loss region.

After penetrating in a certain depth with losing energy by the electron excitation or ionization (LET region), nuclear collisions of the incident ions and the target atoms

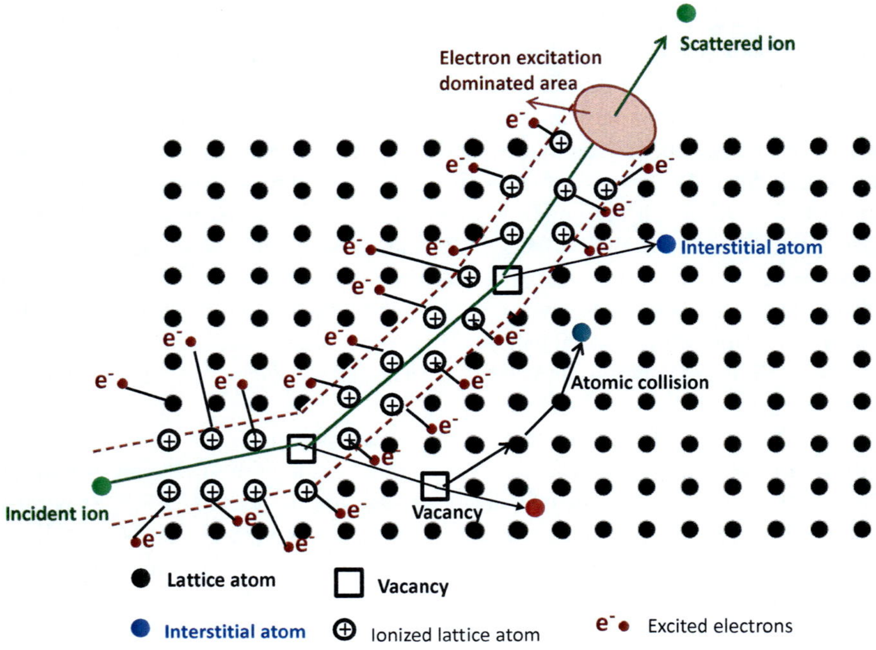

Fig. 4.3 Schematic drawing for energy loss process of energetic particle injected in a solid

become dominant. Because energy loss by the nuclear collision is much larger than that of the ionization, the incident ion stops after repeating the nuclear collision with a much shorter length than the depth of the LET region. The normal depth where the incident ion is stopped from the surface is referred to as a projected rang which is much shorter than the integrated path length of the incident ion until stopped. Figure 4.4 shows depth profiles of deposited energies by the electron excitation, the nuclear collision, and the total loss.

All these collisions are dominated under surface resulting in electron excitation and formation of vacancies and interstitial atoms (referred to as a Frenkel pair for single vacancy and interstitial atom). Most of the exited or released electrons are immediately deexcited or captured by ionized ions releasing energy as photons including X-rays and phonons or heat. Similarly, most of the interstitial atoms and vacancies are recombined. However, some interstitials and vacancies remain making clusters such as vacancy clusters or voids and interstitial loops, referred to as lattice defects which degrade materials properties, hardening, loss of ductility, loss of thermal conductivity, and so on. The formation processes of various radiation damages by energetic ion injection are out of scope of this book and not discussed here. Nevertheless, they strongly influence PMI, in particular, hydrogen recycling and retention which is discussed in Chaps. 6, 8, and 9.

Those energy loss processes that occurred at or near surface appear as PMI and details are described in the following.

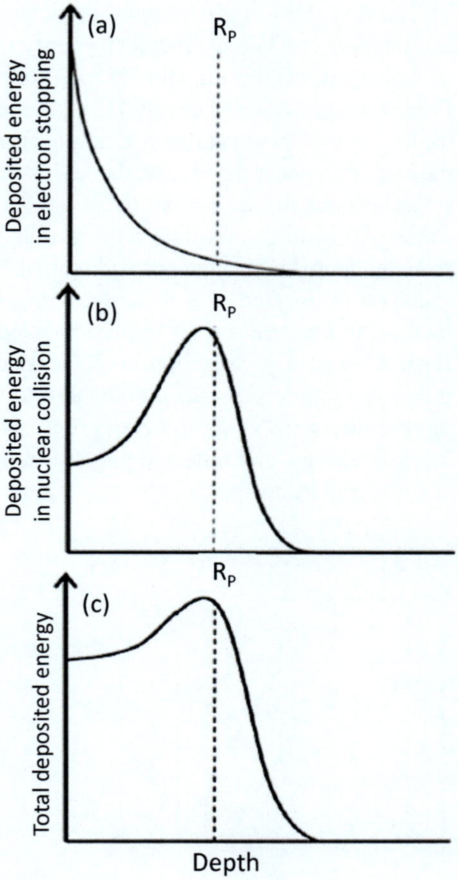

Fig. 4.4 Energy deposition profiles in depth given energetic particle injection; **a** electron stopping, **b** nuclear stopping, and **c** total energy deposition. R_P: Projected range

4.3 Emission of Ions and Neutrals

4.3.1 Reflection

Some of the incident ions are reflected at the first collision with surface constituent atoms without fully losing the incident energy. If the incident ions have sufficient energy, some of them come back to the surface after secondary and tertial collisions. In the case of hydrogen injection, after the incident hydrogen stopping in the target, they diffuse back to the surface, which is referred to as reemission. The difference between the reflection and the reemission is caused by energy carried by released ones, i.e. reflected ones have much high energy than the target temperature, while reemitted ones are mostly thermalized as the target temperature. The reemission is discussed separately in Sect. 4.6.

The flux ratio of all reflected particles to the incident particles is referred to as a reflection coefficient. Incident energy dependence of the reflection coefficients of hydrogen, deuterium, and He ions for carbon (C) and Tungsten (W) is given in Figs. 4.5 and 4.6, respectively [12]. As seen in the figures, the heavier the material (or higher the atomic number (Z number)), the larger the reflection coefficient is, and the larger the incident energy, the less is the coefficient.

Reflected particles are not necessarily ions but include neutrals mostly in excited states. Accordingly, particle reflection accompanies photon emission given by the reflected atoms in excited states. Figure 4.7 shows the Balmer series emissions from reflected D injected in a Si target with different incident energies [13]. With the increasing incident energy, emission intensities deceased, because, as depicted in Figs. 4.5 and 4.6, reflection coefficients decrease. The reduction of the emission intensity is more significant compared to that of the reflection coefficient, because the number ratio of the reflected ions and neutrals decreases with the increasing incident energy, and reflected particles retaining larger energy are in higher excited states or fully ionized states.

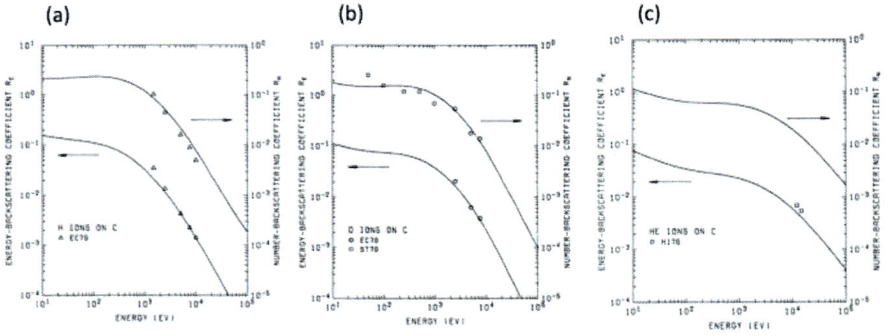

Fig. 4.5 Reflection (backscattering) coefficients of energy and particles (R_E and R_N) for energetic ion injection for **a** H, **b** D, and **c** He to carbon (**c**) [12]

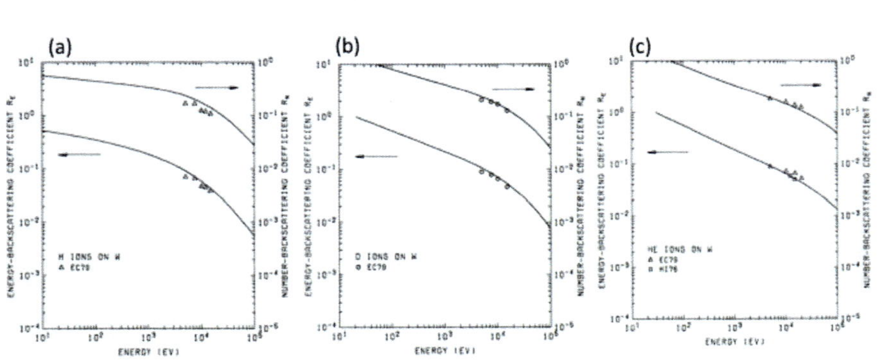

Fig. 4.6 Reflection (backscattering) coefficients of energy and particles (R_E and R_N) for energetic ion injection of **a** H, **b** D, and **c** He to tungsten (W) [12]

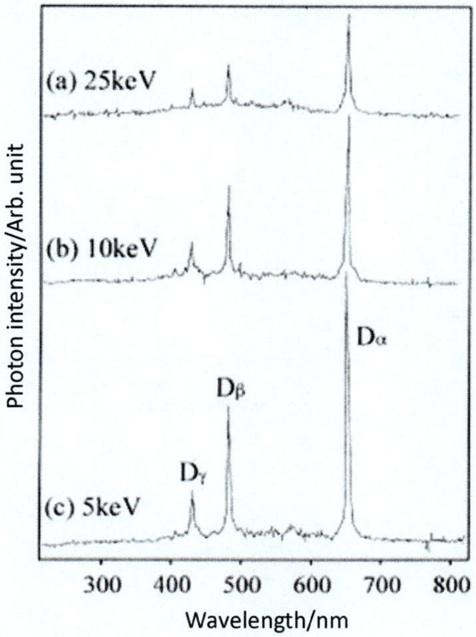

Fig. 4.7 Observed photon emissions from backscattered atoms under irradiation of D^+ ions to clean Si surface for incident energy of **a** 25, **b** 10, and **c** 5 keV. Only the Balmer series emissions of Dα, Dβ, Dγ are appreciable [13]

The Balmer series emission is used as a diagnostic tool to determine the recycling flux of hydrogen between hydrogen plasma and PFS. Figure 4.8a is an example of the Balmer lines emission of D in front of W surface exposed to D plasma [14]. For consideration of the recycling flux, the emission of reflected atoms is to be neglected. However, from detailed measurements of the emissions with changing materials and material temperatures, the emission under plasma exposure was found to include those from the reflected particle. Figure 4.8b is an example of Dα emission from reflected atoms in under 20 keV D^+ injection into Mo, observed using a very fine resolution spectrometer [15]. The observation is a clear indication of the existence of excited atoms in reflected particles. The peak intensity increases with the incident angle corresponding to the increase of the reflection coefficient. Furthermore, peak shift or appearance of a new peak at a shorter wavelength indicates that the velocity of reflected particles increases with the decrease of the incident angle.

Thus, reflected atoms contribute to the Balmer series emission in boundary plasma, and the difference in the reflection coefficients among PFM should have a certain contribution on PMI. Recent observation in ITER-like wall experiments in JET has indicated a large difference in plasma behavior and PMI in comparison with previous JET experiments with full C-wall [16, 17]. The difference in hydrogen recycling properties between the ITER-like wall and the C-wall is also appreciable as discussed in Chap. 8.

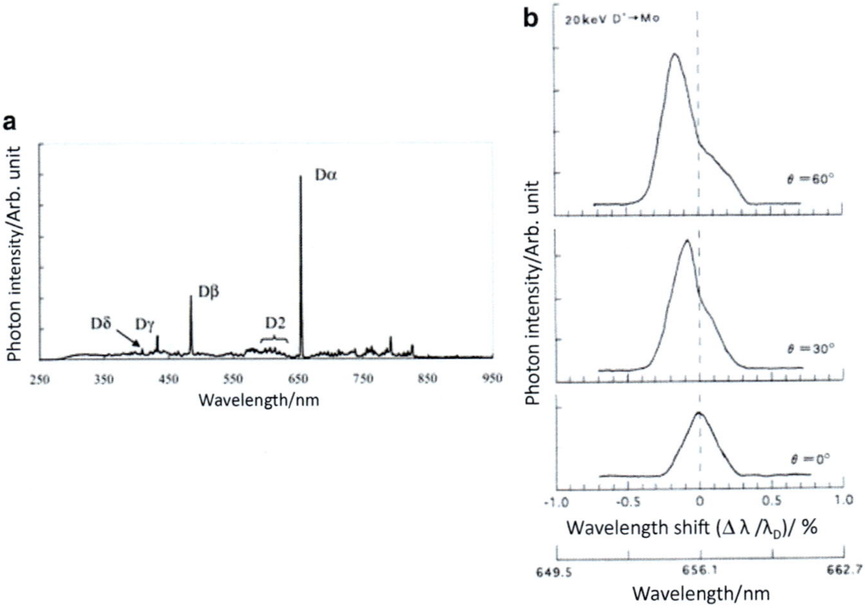

Fig. 4.8 **a** Emission spectra observed in front of W target exposed to deuterium plasma. The Balmer series emission (Dα, Dβ, Dγ, and Dδ), D_2 band spectrum, and the spectrum from the impurities were observed [14]. **b** Changes of Dα spectra emitted from reflected atoms with incident angle under 20 keV D^+ injection on Mo [15]

4.3.2 Physical Sputtering

Early PMI studies focused on sputtering of candidate PFM. Since plasma temperature or ion energy in the plasma was not so high to sputter plasma-facing materials, sputtering studies were mainly done using high-energy ion beams concentrating to understand the physics of sputtering. Nevertheless, plasma confinement (temperature and density) was very poor, mostly caused by the desorption of water vapor and hydrocarbons on PFS. Energetic hydrogen could enhance the desorption and reduction of surface oxide. Therefore, in the early stage of plasma confinement studies, significant effort was put to remove oxygen included in the plasma, which is discussed in this chapter in detail.

Owing to extensive studies, physics of the sputtering is well understood. Except for very low energy region, the binary collision model fits well with the experimental observations, and accordingly simulation codes, such as SRIM [18] and A-CAT [19], are developed and widely used to obtain the dependence of incident energy and angle of primary ion, angular distributions, and energy distributions of sputtered ions and neutrals.

PFS are exposed to energetic particles escaping from plasma as well as residual gas mostly composed of fuels (D_2, DT, and T_2). When the energy of the impinging particles exceeds threshold energy to displace surface atoms, the surface atoms are sputtered to leave the surface, which is referred to as physical sputtering. The threshold energy is larger for heavier surface atoms and the sputtering yield is larger for heavier impinging particles. Since plasma always contains impurities like carbon (C), oxygen (O), and helium (He) produced by D-T reactions, the physical sputtering by these impurities is significantly larger than that by fuel ions. In Fig. 4.9 is compared the energy dependence of physical sputtering yields of W by different kinds of ions [20]. One can clearly see the existence of a threshold energy in the physical sputtering which is larger for higher Z materials. The threshold energy of W sputtering by T ions is above 100 eV, which is one of the reasons to use W as PFM of divertor in a fusion reactor. However, lower Z gases like N, Ne, and Ar injection for edge cooling in a reactor would significantly reduce the threshold as seen in Fig. 4.9. Therefore, the sputtering by the impurities is the main cause of both erosion of PFS and plasma contamination by sputtered atoms. Physical sputtering yields change with the angle of incident ions, showing higher values for lower incident angles; as expected nuclear stopping region becomes shallower for low angle incidence.

The sputtered atoms are mostly neutral showing the Maxwell–Boltzmann-like energy distribution with a maximum energy of a few eV. In boundary plasma (or scrape-off layers) of a reactor, released atoms from PFM by sputtering are immediately ionized and succeedingly gyrated along magnetic field lines. Consequently, most of them return to be deposited nearby the sputtered location (referred to as prompt deposition) or transferred along the scrape-off layers to be deposited somewhere on the plasma-facing surface, shadowed area, and/or exhausted from the reactor. Some of the ionized atoms penetrate deep into plasma and are confined as impurity ions. Repetitive processes of erosion by sputtering and deposition cause

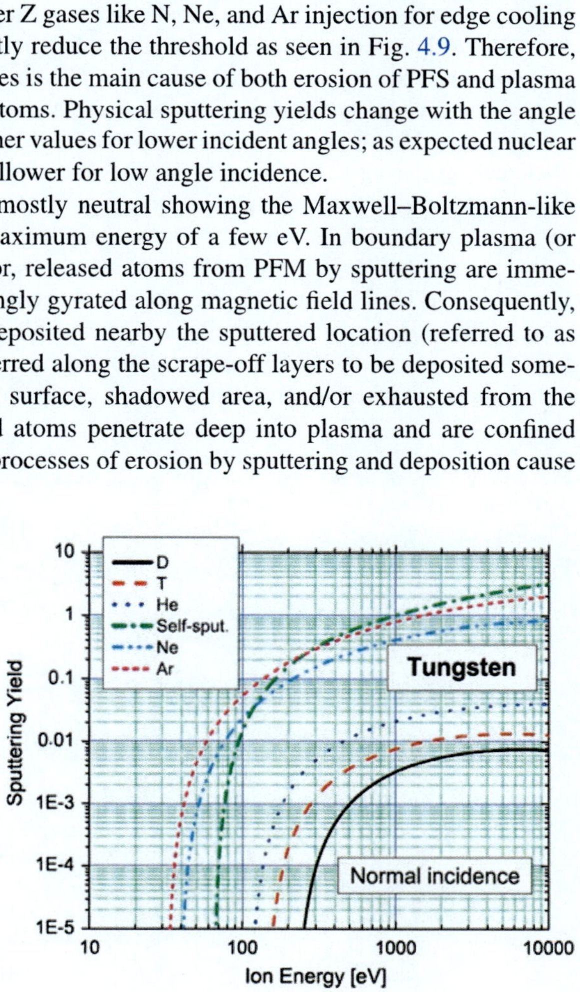

Fig. 4.9 Sputtering yields for tungsten by hydrogen and some impurities at normal incidence angle [20]

long-range transport of eroded materials. Those deposited at plasma shadowed areas, like gaps of plasma-facing tiles and vacuum ducts, are simply piled-up without re-sputtering.

In the deposited layers, fuel particles are easily incorporated, often referred to as co-deposition. Therefore, the incorporation of the fuel in the deposited materials at the plasma shadowed area becomes the dominant cause of T retention in the reactor vessel (in-vessel T inventory).

4.3.3 Chemical Sputtering

When plasma-facing wall is made of materials reactive to hydrogen, like C and Beryllium (Be), the material surface can be eroded by chemical reaction with hydrogen-producing volatile products, like CH_4 and BeH_2 (chemical sputtering) additionally to physical sputtering. Figure 4.10 shows the incident energy dependence of carbon sputtering by H and D ions with the separation of the physical sputtering and the chemical sputtering [21]. One can see that the chemical sputtering dominates at lower incident energies without the sputtering thresholds. Since chemical reactions between H and C change with temperature, the chemical sputtering yields of C show strong temperature dependence with its maximum yield at around 800 K as shown in Fig. 4.11 [22]. This is because the reaction rate to produce CH_4 increases with temperature, while CH_4 becomes unstable at higher temperatures. The decomposition of CH_4 is used to manufacture pyrolytic-carbon and diamond films. Above 800 K, owing to radiation-enhanced sublimation which is typical for carbon materials, the erosion

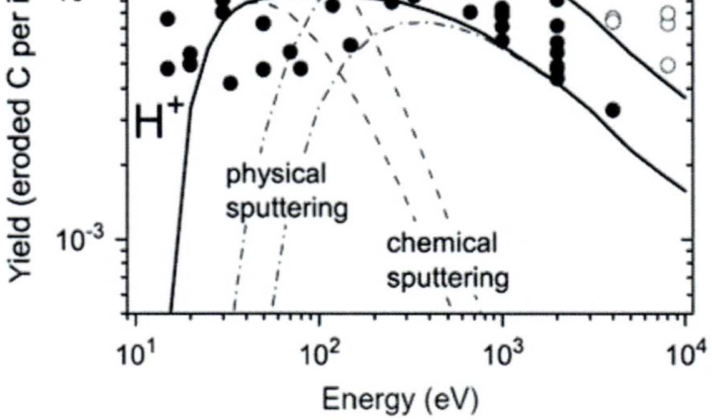

Fig. 4.10 Sputtering yields for carbon by H^+ and D^+ ions (reprinted with permission from [21])

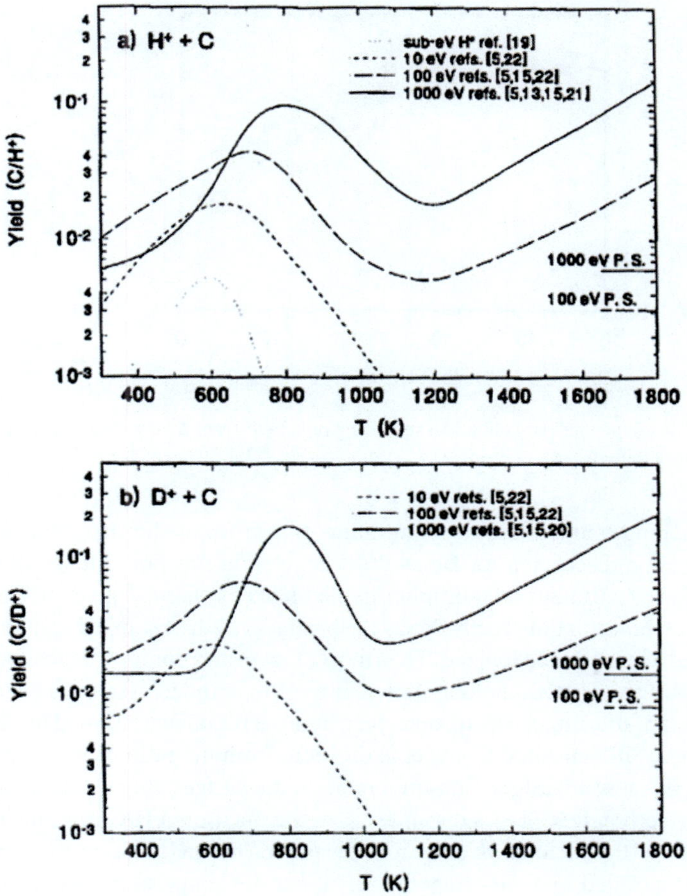

Fig. 4.11 Temperature dependence of carbon sputtering by H^+ and D^+ ions (reprinted with permission from [22])

yield increases again. The radiation-enhanced sublimation is separately discussed in the next section. Different from the physical sputtering, the chemical sputtering of C shows flux dependence as shown in Fig. 4.12 [23]. With the increasing ion flux, the chemical sputtering yields significantly decrease, and in reactor condition with the flux of around 10^{24} m^{-2} s^{-1}, more than one order of magnitude less than those for lower flux. The reason for the reduction is not fully understood, yet.

The chemical stability of the C-H bond is also the cause of the H incorporation in deposited C layers of high H retention. Similar to C, Be is chemically sputtered by H plasma making BeH_2 [24] and erosion of Be is not small. Correspondingly, deposited Be layers retain a large amount of H making BeH_2. Erosion and deposition of Be are quite important in JET [25] because Be is used as the first wall. However,

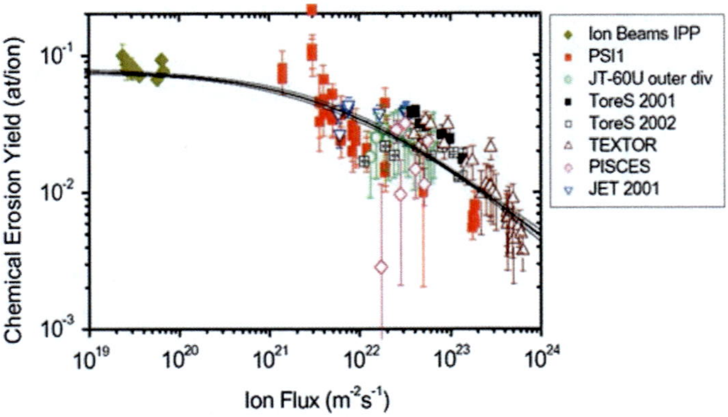

Fig. 4.12 Flux dependence of chemical sputtering of C by H (reprinted with permission from [23]) See references therein

its low melting temperature does not allow to use Be as the first wall in a fusion reactor, and no discussion for Be as PFM is given in this book. It should be noted that Be is used as a neutron multiplier in the blanket system as given in Chap. 7.

Hydrocarbons, mainly methane, as the products of the chemical sputtering of C, are emitted to plasma and ionized. Then most of them are promptly deposited by gyration in a strong magnetic field at the vicinity of the eroded area to make deposited layers. It is quite important to note that hydrogen concentration (H/C in atomic number ratio) in deposited C layers is different from the ratio of incident fluxes of H and C. Because hydrogen flux impinging to the surface (ϕ_H) is much larger than impinging carbon flux (ϕ_C), i.e. $\phi_H/\phi_C \gg 4$, the maximum H/C in stable hydrocarbons, most of the incident hydrogen is reemitted. Instead, H/C in the deposited C layers is controlled by their temperature; higher the temperature, the lower is H/C. This is the reason to use "deposits and/or deposited layers" but not "co-deposits and/or co-deposited layers" in this book. If H/C exceeds around 1, the layers become volatile, resulting in the chemical sputtering.

In present tokamaks, as shown in Fig. 5.1, most of the first wall and the outer divertor are eroded and the inner divertor deposited. Repetitive processes of erosion and deposition transport carbon eroded at the outer divertor and the first wall mostly to the inner divertor. At plasma shadowed area, such as gaps of armor tiles and pumping ducts, the deposited C layers are not subjected to plasma and continue to grow in their thickness. Large erosion by chemical sputtering and large T retention in the deposited C layers are the main reason to exclude carbon materials as its PFM in ITER.

4.3.4 *Ion-induced Desorption and Radiation-Enhanced Sublimation*

Incident particles deposit their energy to surface constituent atoms so as impurity atoms and molecules adsorbed on the surface. Ion-induced desorption is observed when surface adsorbed or absorbed atoms or molecules get enough energy to desorb from the surface. Although the energy of the released particles is lower than that of sputtered atoms, it is much larger than that of the thermalized one at the surface. The particle flux caused by the ion-induced desorption is much higher than thermally desorbed ones and released particle flux becomes much larger than that expected from the vapor pressure of the target material, which is referred to as radiation-enhanced sublimation (RES). RES occurs in any materials and becomes appreciable far below their melting temperature. In particular, carbon materials show large RES above 1200 K as seen in Fig. 4.11. For utilization of C as PFM at elevated temperatures, RES should be taken into account.

Plasma-induced desorption or discharge-induced desorption of adsorbed molecules typically water and hydrocarbons in a vacuum vessel is appreciable as strong impurity radiation during the plasma ramp-up process, often resulting in plasma collapse. The surface cleaning before plasma discharge has been very important.

4.4 Emission of Electrons and Photons

4.4.1 *Electron Emission*

Energetic ion injection induces electron emission often referred to as secondary electron emission. Because of the secondary electron emission from the target under the ion injection, target current measurements to determine the incident ion flux often result in overestimation. That is the reason to use the Faraday cup which suppresses the secondary electron emission for the determination of the injected ion current. If the surface is contaminated with some absorbents, the secondary electron emission is more appreciable, and measured current often becomes two or three times large than the incident ion current.

For secondary electron emission, there are several different mechanisms. They are ion-induced kinetics emission, potential emission caused by ion injection, photoinduced emission by photon injection, and thermionic emission caused by the target temperature rise.

The ion-induced kinetic emission is caused by the collision of the incident ion (or its belonging electrons) with electrons of constituent elements of the target. The ion-induced Auger electron emission is included in the kinetic emission. The potential emission is caused by a large electric field induced between the incident ion and the target surface. Figure 4.13 shows the energy distribution of secondary electrons

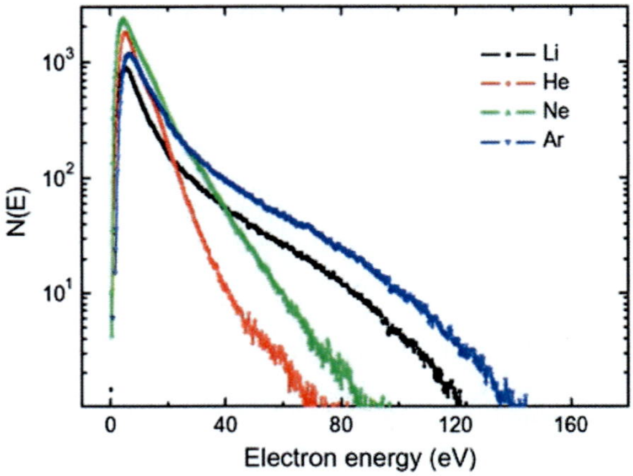

Fig. 4.13 Energy distribution of secondary electrons induced by injection of Li, He, Ne, and Ar ions on Cu(100) for 5 keV incident energy. Spectra are corrected by the energy transmission of the analyzer (reprinted with permission from [26])

induced by injection of Li, He, Ne, and Ar ions on Cu(100) for 5 keV incident energy [26]. Heavier the incident ion, the secondary electrons with higher energy increase, because interactions between the incident ion and the target occur in the shallower region resulting in less energy loss of the secondary electrons. Together with the kinetic electron emissions, X-rays or photons are emitted including characteristic X-rays (like the Auger electron emission).

Generally, energetic electron injection gives similar phenomena to those given by energetic particle injection. They are reflection; emission of secondary particles including secondary electrons, photons, and phonons; sputtering of constituent elements of target materials; and electron-induced desorption. Because of the much lower mass of an electron, however, these emissions are less appreciable compared to ion-induced emissions. Nevertheless, huge incident electron flux accompanying huge ion flux at PFS could have a significant influence on PMI in a reactor. Until now the effect of the electron has not been considered or examined because electron flux in current tokamak is not so high as that in a reactor.

In tokamak discharges, runaway electrons of which energy sometimes exceed MeV appear. Owing to their very high energy, material damages like arching and hot spot and resultant local melting give a significant influence on materials integrity.

4.4.2 Photon Emission

As described above, energetic particle injection always accompanies emissions of photons including X-rays, UV, Visual light, and even infrared radiation by the temperature rise of the target. Detection of characteristic X-rays under high-energy proton is used as an analysis technique of elements in materials referred to as PIXE (Proton-Induced X-ray Emission). The photoemission could induce photoelectrons. The thermionic emission appears if deposited energy by injected ions is so large that the target temperature becomes quite high.

The radiation or emitted photons can be used as a diagnostic tool to measure surface temperature. However, since radiation power is proportional to the fourth power of absolute temperature, the radiation becomes appreciable only when the target temperature increases over around 2000 K. In a divertor target area where power load by plasma particles is very large, radiated power from the target could have some influence on power valance. Transition heat load such as ELM and arching makes the local hot spot which can be detected from the radiation. The hot spot enhances photoelectron emission which, in turn, attracts more incident ions resulting in temperature escalation to make the hot spots [27].

4.5 Energy Reflection

As already described above, part of energetic particles and photons incident to the material surface are reflected with energy distributed from the same energy of incident to zero depending on the incident energy, incident angle, and target materials. As examples, reflection of ions and energy carried by reflected ions are given in Figs. 4.5 and 4.6, respectively for C and W, as reflection coefficients of ion and energy for incident ions of H, D, and He. In energy reflection, not only energy carried by primary ions and photons but also energy carried by secondary particles and photons are included. Furthermore, as discussed in the previous section, radiation from the target materials becomes high if deposited energy by primary particles and photons are high enough to make the target temperature very high. As described in the previous section, the radiation from the target material becomes appreciable above around 2000 K. In the divertor target area, owing to very high heat load, the effect of energy reflection and radiation from the target materials would become significant. In material power load tests using electron or ion beam, the energy reflection is taken into account.

4.6 Reemission of Incident Ions

Initially injected ions except reflected ones are retained in a target material. Accordingly, the concentration of the injected ions in the target increases. When their concentration in the subsurface region becomes the saturation concentration, reemission of the incident ions starts. After long implantation, although the depth of saturated region gradually broadened, most of the injected ions are reemitted, in other words, reemitted flux becomes nearly the same as the incident flux. Different from the reflected ones, reemitted particles are mostly thermalized at the target temperature. Reemission behavior is quite different between fuels and inert gases such as He produced by fusion and Ne and Ar seeded for edge cooling. In the following sections, they are discussed separately.

4.6.1 Reemission of Hydrogen (Fuel)

In the case of fuel (referred to as hydrogen hereafter), some of implanted hydrogen diffuse into bulk and reemission delays, even permeates through. Figure 4.14 is an example showing how reemission occurs under energetic D ion injection to 316 SS at −120 °C [28]. At the start of the ion injection, the contribution of reflected D ions which are recombined to molecules at the system wall dominates then reemission of recombined molecules at the target surface increase with time and finally, reemission flux is balanced to the incident flux showing steady reemission. For higher incident energy, the reflected fraction is lower and the time to attain the steady reemission becomes longer, because of deeper injection.

For hydrogen injection, reemission is not necessarily as hydrogen molecules. When temperature of the target surface is over 1200 K, atomic hydrogen without recombining to molecules is reemitted as given in Fig. 4.15 [29]. Although until now there are no tokamaks having metallic divertor with temperature over 1200 K, in ITER or a reactor which use W divertor, the effect of atomic reemission would be appreciable. Without recombination, energy dissipations by the recombination do not occur. Accordingly, energy reflection at the surface would be enhanced, which is beneficial for reducing power load and must be confirmed in ITER.

Because most of the metal surface is contaminated by water absorption or oxidized, water molecules are thermally or collisionally desorbed by injected hydrogen ions as ion-induced desorption and reemitted hydrogen also reduce the surface oxide to release H_2O molecules. Figure 4.16 shows how reemitted D_2 molecules change with time under some oxygen-containing vacuum and for the oxidized surface. Since part of reemitted D reacts with O on the surface [30], reemission as D_2 is lower compared with that for clean surface until surface oxygen is fully removed, while under some oxygen-containing atmosphere, production of D_2O at the surface continues to give steady-state reemission lower than that for the clean surface.

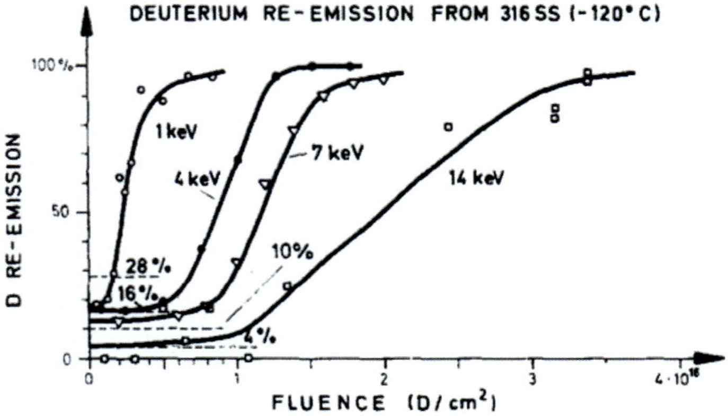

Fig. 4.14 Deuterium reemission rate for 316 stainless steel injected with D^+ ions having different energy at −120 °C normalized with the incident flux (reprinted with permission from [28])

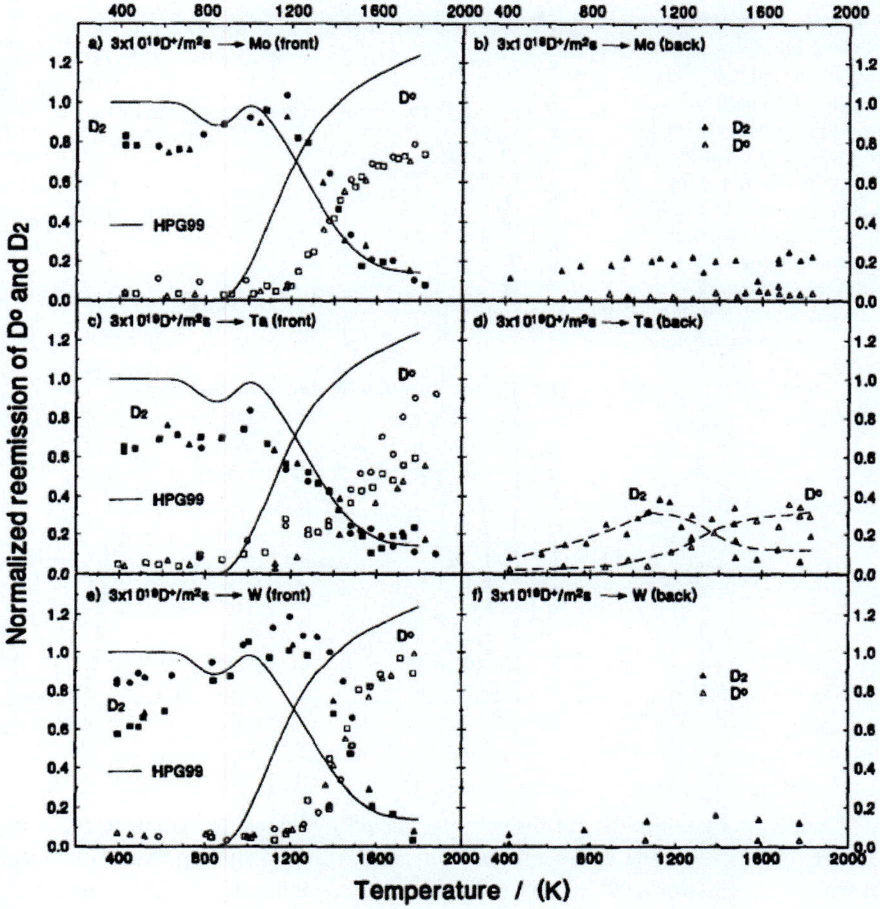

Fig. 4.15 Steady-state reemission of atoms and molecules from the front surface and the back surface as a function of sample temperature for Mo (**a**, **b**), Ta, (**c**, **d**), and W (**e**, **f**) under irradiation of 3 keV D_3^+ ions (i.e. 1 keV/D^+) incident on the sample at an angle of 35° with respect to the sample normal. The reemission data were normalized by setting a single normalization constant for all three specimens based on the average value of the molecular reemissions at ~1000 K; thus the normalized value of the sum of the molecular reemission from both surfaces at 1000 K is 1 ± 0.1. Solid symbols are for D_2 and open symbols are for D° (D atom). Different symbol shapes represent data taken on different days, in order of open then closed circles, open then closed squares, and open then closed triangles. The cumulative fluence associated with each data set is approximately (2–3) $\times$ 10^{23} D m^{-2} The solid lines correspond to reemission results for pyrolytic graphite (reprinted with permission from [29])

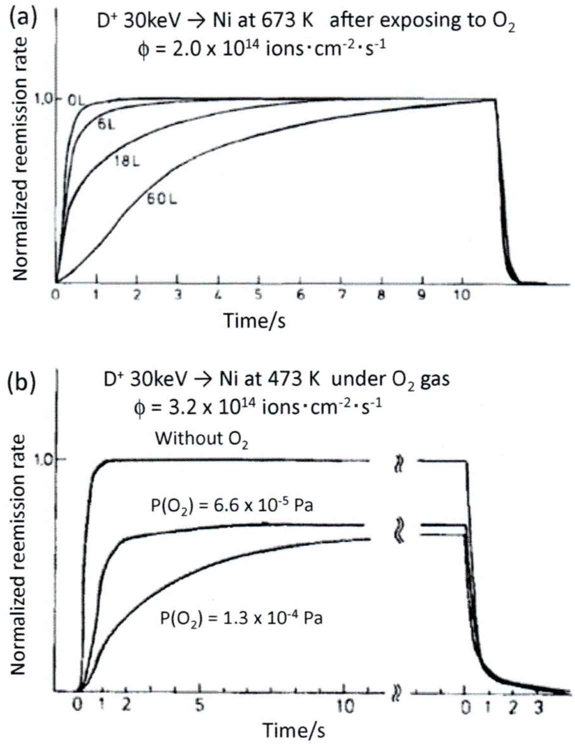

Fig. 4.16 **a** Time sequences of deuterium reemission rate for Ni injected with 30 keV D^+ ions. Before the injection, Ni was exposed to oxygen gas at 1.3×10^{-5} Pa. Exposure is given in the Langmuir units, L (1 L = 1.3 $\times 10^{-4}$ Pa s^{-1}). **b** Time sequences of deuterium reemission rate for Ni injected with 30 keV D^+ ions under oxygen atmosphere [30]

Chemical erosion is the reemission of injected hydrogen in carbon in the form of CH_4 or other volatile hydrocarbons. Different from the surface oxide, chemical erosion or reemission with the form of hydrocarbon continues and stays constant.

In a fusion reactor, the incident flux of hydrogen is so large that surface contaminants would be immediately removed if they make volatile molecules. Hence, only the ramp-up phase of the plasma would be influenced by released molecules like water and hydrocarbons. As seen in the energy dependence of reflection coefficients, the reflected fraction would dominate the reemission in boundary plasma.

4.6.2 Reemission of Inert Gas Atoms

Inert gas injection results in the formation of bubbles by the accumulation of injected atoms in injected zones and the bubbles coalesce to blisters. Accordingly, the geometrical structure is significantly modified. A specific structure referred to as fuzz appears on W surface with exposure of low-energy He ions at a higher temperature range. Because D-T reactions produced He, injection of He to PFS could result in the fuzz formation and its influence on PMI are concerned. Figure 4.17 is an example of the

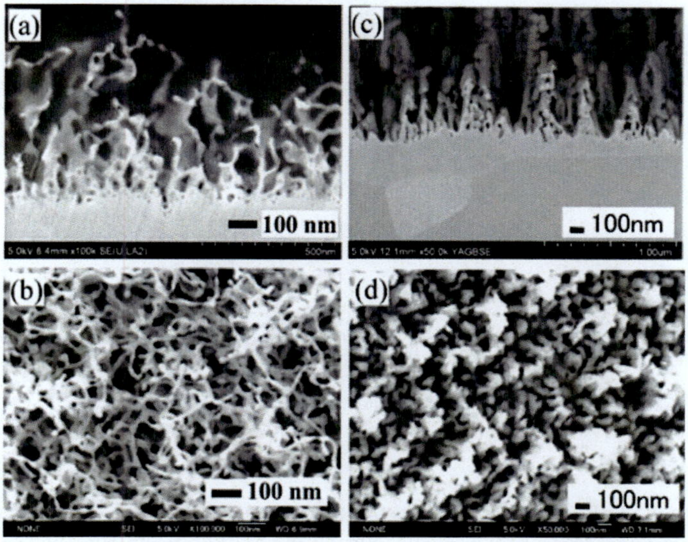

Fig. 4.17 Fiber-form nanostructure (fuzz structure) of PM-W tungsten with complete black color obtained by He plasma exposure in AIT-PID. **a** and **b** show the top surfaces with different scanning electron microscopy (SEM) magnifications, **c** shows a cross section, and **d** is a photo showing surface blackening (reprinted with permission from [31])

fuzz on W surface produced He plasma exposure [31]. Various papers have been published on the formation of the fuzz and its effects on PSI. On high power loaded area, even if the fuzz was formed, its very fine structure would not be kept owing to material softening, melting, or evaporation. Nevertheless, some events, such as arching, evaporation, and melting are concerned.

Because the fuzz appears at a rather high temperature where vacancies or He bubbles induced by He injection move, the formation process correlates with He migration or He reemission from the surface. The temperature range for fuzz formation differs from metal to metal and can be correlated with the melting point of metals.

Figure 4.18 shows He reemission injected in Aluminum (Al) at RT. In addition to small burst-like emission, i.e. larger reemission flux compared to the incident flux appeared several times [32]. The small burst and larger one are likely corresponding to the fracture of small bubbles or blisters and fracture of large areas, respectively. Figure 4.19 shows He reemission observed of He injection in Ni at various temperatures [32]. Periodic large emissions correspond to accumulated burst-like emission resulting in exfoliation of a large area. After one large emission, He accumulation occurs again and repeating periodic burst-like reemissions. He movement becomes smoother with increasing temperature, and steady-state reemission balanced with the incident flux is attained after several large burst-like emissions. Above 840 K, no more periodic emission appears and reemission becomes smooth like H reemission. Figure 4.20 summarizes changes of He reemission behavior against normalized

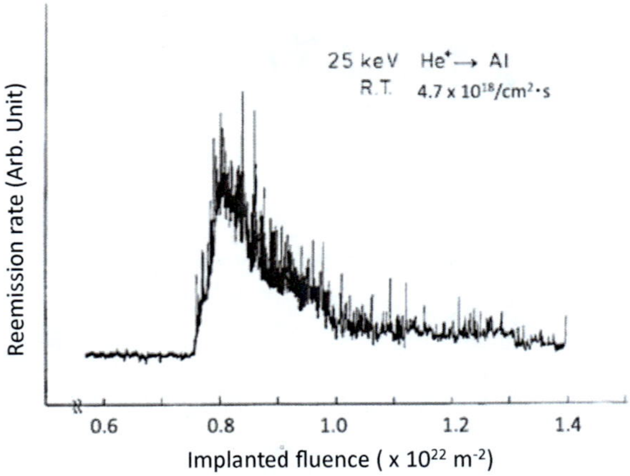

Fig. 4.18 Reemission of He implanted in Al at RT with the incident energy of 20 keV and flux of 4.7 $\times$ 10^{14} cm^{-2} s^{-1} [32]

temperatures with melting points of target materials [32]. In the temperature region assigned as range III in Fig. 4.20 fuzz structure should appear. Because of the large size of He atoms, its migration is like vacancy migration accompanying movements of substrate atoms suppressing the formation of large size bubbles.

Under the surface regions where energetic He ions or other ions are injected, various defects are formed such as interstitials, vacancies, their clusters, bubbles, dislocation loops, and so on, which have been extensively studied as radiation damages causing degradation of materials properties but are not discussed in this book.

4.7 Interaction of Released Particles with Photons and Electrons in Boundary Plasmas

Interactions of released particles from PFS with electrons and photons in boundary plasma have been the main subject of PMI. Various phenomena appear in boundary plasma and have been extensively studied. Interactions of C released from divertor with boundary plasmas in carbon machines have been quite important in C-wall tokamak and extensively studied. Figure 4.21 is a good example showing ionization and recombination of C atom or ions with the impact of electron/photon [33]. For the C-wall, the amount of eroded materials (C and hydrocarbon) is large. Although most of the eroded materials from PFS are promptly deposited, as discussed above, appreciable amounts of them enter boundary plasma and ionized as indicated in Fig. 4.21. The ionization takes energy from boundary plasma and contributes to its cooling. Subsequently, the ionized C captures electrons referred to as recombination, which releases energy as radiation contributing to the cooling of the boundary plasma. In recombination plasma with high density neutral in divertor, the power exhaust driven by the recombination process is so powerful and dominate the PMI process, as often observed as MARFE in JT-60U [33] and Carbon blooms in JET and TFTR [34].

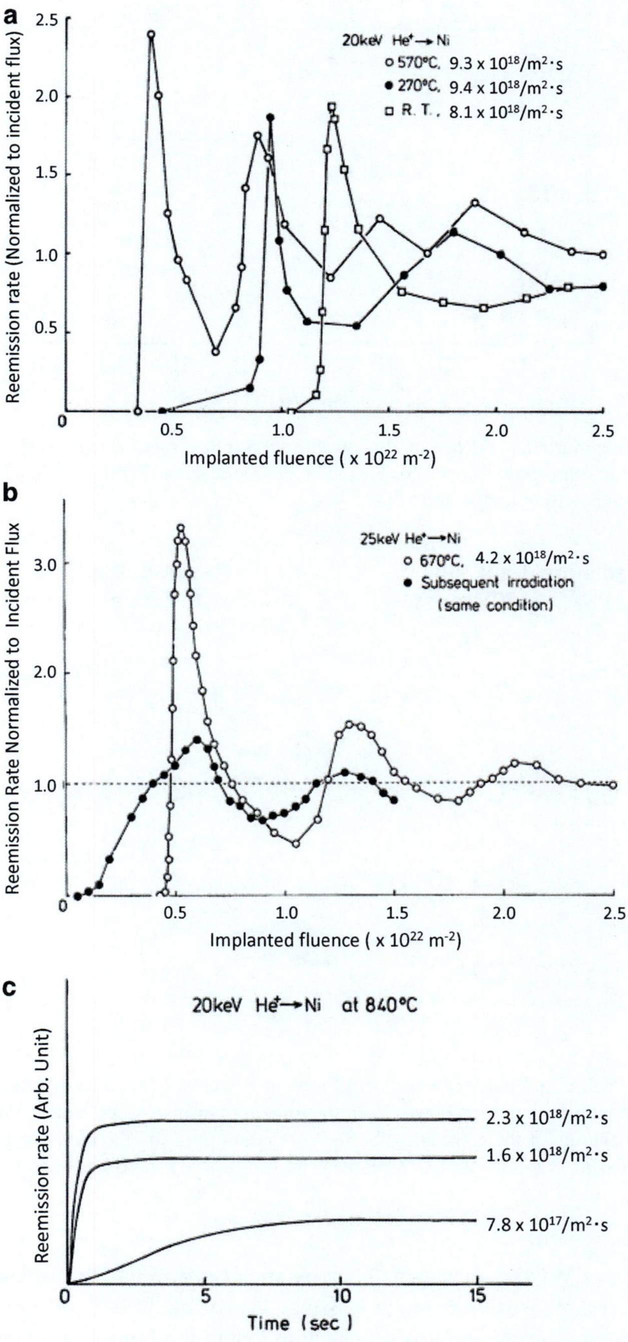

Fig. 4.19 **a** He reemission implanted in Ni with the incident energy of 20 keV at RT, 270 °C and 570 °C, **b** at 600 °C and 780 °C with the incident energy of 25 keV, and **c** at 840 °C with the incident energy of 20 keV for three different fluxes [32]

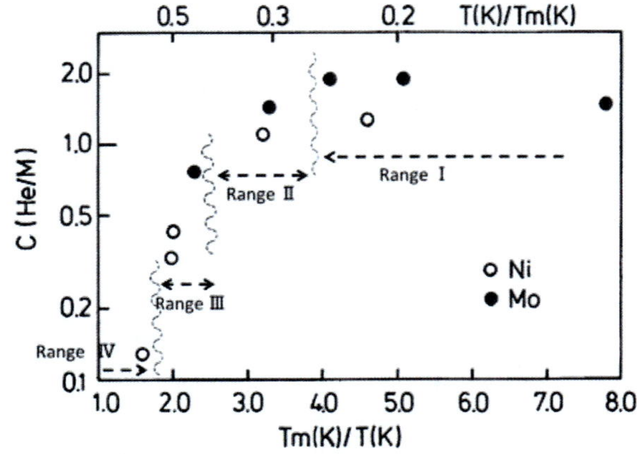

Fig. 4.20 Temperature dependence of He concentration in the injected region of Ni and Mo with 20 keV He^+ at the threshold fluence for burst-like He reemission (See Fig. 4.19). Temperature is normalized to the melting temperature [32]

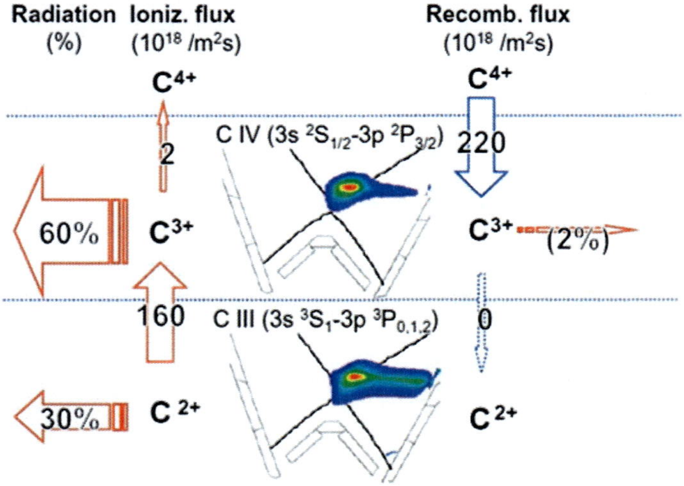

Fig. 4.21 Ionization and recombination of C ions in detached plasma in divertor region of JT-60U. From the left, the radiation power from the ionizing plasma component, the ionization flux, the spatial distribution of the emissivity, the recombination flux, and the radiation power from the recombining plasma component of C^{3+} (upper) and C^{2+} (lower) (reprinted with permission from [33])

In the case of W-wall, energy exhaust by such recombination process should not be large because W erosion is much less than C erosion. To enhance power exhaust in the divertor area, some impurity gases with low to medium Z number are planned to be seeded intentionally in the boundary plasma in ITER and a reactor (impurity seeding).

As such, atomic and molecular processes in boundary plasma have been the main subjects in PMI studies and various reviews and textbooks have been published, for example, see references [5, 35–37]. In this book, details of these atomic and molecular processes in boundary plasma are not discussed.

4.8 Summary

Released particles from PFM into boundary plasma include three different ones, i.e. fuels or H, sputtered particles and other impurities released from PFM, and electrons. In addition, photons are directly emitted from PFS. Since plasma discharge prefers wall pumping, H recycling was not systematically studied except Hα measurement which has been used to determine hydrogen influx. However, the amount of retained fuel in PFM must be far larger than that in the burning plasma, some changes in the fuel retention caused by temperature rise given by localized or transient power load would influence plasma confinement. Therefore, the dynamic behavior of hydrogen in PFM should play a quite important role. This chapter is devoted to foresee it in a reactor based on fundamental physics and chemistry of hydrogen behavior in candidate PFM.

Erosion is also quite important in respect of lifetime of PFM and H retention in deposited materials. Since the incident fluence of fuels in present tokamaks is still far less compared to that of a reactor, no PFM was changed owing to erosion, and H retention in deposited layers in plasma shadowed area gave no significant influence on plasma. Therefore, little attention has been paid to erosion and deposition and accompanied lifetime of PFM in present tokamaks. However, the lifetime of PFM in a reactor should be estimated prior to use. In addition, usage of T as a fuel requires strict control in its inventory for both T safety and fuel self-sufficiency. The subjects written here would help for both.

Most of the PMI studies up to now are relating the interaction of released particles from PFM and boundary plasmas, and analysis of photons from the boundary plasma has been the main target in PMI studies. Because spectroscopy has been the most important diagnostic, we have to continue to rely on it. In a fusion reactor, available ports for diagnostics would be limited, spectroscopy and other diagnostic techniques shall be selected appropriately to be useful for controlling the burning plasma. For that, this chapter would give important information.

References

1. B.D. Wirth, K.D. Hammond, S.I. Krasheninnikov, D. Maroudas, Challenges and opportunities of modeling plasma–surface interactions in tungsten using high-performance computing. J. Nucl. Mater. **463**, 30–38 (2015)
2. *Radiation effects on Solid surfaces*, Ed. M. Kaminsky, American Chemical Soc., Washington (1976). ISBN: 0-8412-0331-8

3. P.D. Townsend, J.C. Kelly, N.E.W. Hartley, *Ion Implantation, Sputtering and their Applications.* Academic Press (1976). ISBN: 0-12-696950-7
4. W.O. Hofer, J. Roth, *Physical Processes of the Interaction of Fusion Plasmas With Solids.* Academic Press (1996). ISBN: 0-12-351530-0
5. R.K. Janef, H.W. Drawin, *Atomic and Plasma-Material Interaction Processes in Controlled Thermonuclear Fusion*, Ed. Elsevier 1993: ISBN: 0-444-81630-5
6. D. Reiter, M. Baelmans, P. Boerner, The EIRENE and B2-EIRENE Codes. Fusion Sci. Technol. **47**, 172–186 (2005)
7. ERO Code webpage. http://www.ero-code.de/ero_old/index.html
8. T. Takizuka, Development of the PARASOL code and full particle simulation of tokamak plasma with an open-field SOL-divertor region using PARASOL. Plasma Sci. Technol. **13**, 316 (2011)
9. H. Kawashima, K. Shimizu, T. Talizuka et al., Development of integrated SOL/divertor code and simulation study in JAEA. Plasma Fusion Res. **1**, 031 (2006)
10. S. Wiesen, D. Reiter, V. Kotov et al., The new SOLPS-ITER code package. J. Nucl. Mater. **463**, 480–484 (2015)
11. Users manual for the UEDGE edge-plasma transport code. https://www.osti.gov/biblio/15007243
12. T. Tabata, R. Ito, Y. Itikawa, et al., Data on the backscattering coefficient of light ions from solids, Rep. IPPJ-AM-18 (1981); R. Ito, T. Tabata, T. Itoh, Data ow the backscattering coefficients of Light ions from Solids, (A Revision) Rep. IPPJAM-41 (1985)
13. A. Ohmori, T. Tanabe, Influence of target chemical activity on Balmer lines emission from backscattered hydrogen. J. Nucl. Mater. **258–263**, 666–671 (1998)
14. K. Shimada, T. Tanabe, R. Causey et al., Hydrogen recycling study by Balmer lines emissions in linear plasma machine TPE. J. Nucl. Mater. **290–293**, 478–481 (2001)
15. T. Tanabe, K. Ohya, N. Otsuki, Hydrogen reflection and H α emission. J. Nucl. Mater. **220–222**, 841–845 (1995)
16. G F Matthews, P Edwards, T Hirai, et al., Overview of the ITER-like wall project Physica Scripta, **T128**, 137–143 (2007)
17. S. Brezinsek, T. Loarer, V. Philipps, et al., Fuel retention studies with the ITER-Like Wall in JET. Nucl. Fusion **53**, 083023 (13 pp) (2013)
18. James F. Ziegler, SRIM—The Stopping and Range of Ions in Matter. http://www.srim.org/
19. Y. Yamamura, Dynamic MC simulation of low-energy ion implantation. Nucl. Instr. Methods **B153**, 410–414 (1999)
20. Y. Yamamura, Y. Itikawa, N. Itoh, Angular dependence of sputtering yields of monoatomic solid, IPPJ-AM-26 (1983), Nagoya University
21. C. Hopf, W. Jacob, Bombardment of graphite with hydrogen isotopes: a model for the energy dependence of the chemical sputtering yield. J. Nucl. Mater. **342**, 141–147 (2005)
22. A.A. Haasz, J.A. Stephens6, E. Vietzkec, et al., Atomic and Plasma-Materials Interaction Data for Fusion, **7A**, 9–63 (1998)
23. J. Roth, A. Kirschner, W. Bohmeyer et al., Flux dependence of carbon erosion and implication for ITER. J. Nucl. Mater. **337–339**, 970 (2005)
24. R.P. Doerner, M.J. Baldwin, D. Buchenauer The role of beryllium deuteride in plasma-beryllium interactions. J. Nucl. Mater. **390–391**, 681–684 (2009)
25. S. Brezinsek, JET-EFDA contributors, Plasma-surface interaction in the Be/W environment: Conclusions drawn from the JET-ILW for ITER. J. Nucl. Mater. **463**, 11–21 (2015)
26. G. Ruano, J. Ferrón, Ion induced high energy electron emission from copper. Nucl. Instr. Methods B **266**, 4888–4890 (2008)
27. I.V., T. Tanabe, The Influence of electron emission on heat load to the plasma facing materials under space charge limited condition with an oblique magnetic field. J. Nucl. Mater. **266–269,** 714–720 (1999)
28. R.S. Blewer, R. Behrisch, B.M.U. Scherzer, R. Schulz, Trapping and replacement of 1–14 keV hydrogen and deuterium in 316 stainless steel. J. Nucl. Mater. **76/77**, 305 (1978)

29. J.W. Davis, A.A. Haasz, Reemission of deuterium atoms from Mo, Ta and W during D^+ irradiation. J. Nucl. Mater. **223**, 312–315 (1995)
30. T. Tanabe, M. Takeo, S. Imoto, Effect of surface oxygen on reemission of deuterium implanted in nickel. J. Nucl. Mater. **185**, 286–291 (1991)
31. S. Takamura, T. Miyamoto, N. Ohno, Effects of fiber-form nanostructures on particle emissions from a tungsten surface in plasmas. Nucl. Fusion **52**, 123001 (2012)
32. T. Tanabe, He reemission implanted in metals. J. Nucl. Mater. **453**, 247–252 (2014)
33. T. Nakano, H. Kubo, N. Asakura et al., Radiation process of carbon ions in JT-60U detached divertor plasmas. J. Nucl. Mater. **390–391**, 255–258 (2009)
34. M. Ulrickson, The JET Team, The TFTR Team, a review of carbon blooms on JET and TFTR. J. Nucl. Mater. **176 & 177**, 44–50 (1990)
35. D.E. Post, A review of recent developments in atomic processes for divertors and edge plasmas. J. Nucl. Mater. **220–222**, 143–157 (1995)
36. G. Janeschitz, ITER JCT and HTs, Plasma-wall interaction issues in ITER, J. Nucl. Mater. **290–293**, 1–11 (2001)
37. R.E.H. Clark, D.H. Reiter (Eds.) Nuclear fusion research Understanding plasma-surface interactions. Springer Series in Chemical Physics, Springer 2010, ISSN 0172-6218, ISBN 987-3-642-06197-4

Chapter 5
Erosion and Deposition, and Their Influences on Plasma Behavior (Material Transport in Tokamak)

5.1 Introduction

Very high photon and particle fluxes to plasma-facing surfaces (PFS) in a fusion reactor cause materials' erosion like physical and chemical sputtering and sublimation. The eroded materials which are mostly neutral particles enter into boundary plasma. Subsequently, they are ionized and gyrated along magnetic field lines. Since in the boundary plasma, the magnetic field lines attach to PFS with tiny contact angles, most of the gyrated ions return to the surface to make deposited layers close to the eroded location. (referred as prompt deposition). Therefore, direct long-range transfer of the eroded materials does not likely occur, but repetitive processes of erosion and deposition transport the eroded materials in long distance as shown in Fig. 5.1 in large tokamaks. In present divertor tokamaks, the inner first wall and the outer divertor are mostly eroded, while the inner divertor is deposited. The outer first wall is either eroded or deposited depending on location. In tile gaps particularly for surface eroded tiles, some of the eroded materials penetrate to make deposited layers. In carbon wall tokamaks, significant deposition appears on locations not directly facing to plasma but on the line of sight such as divertor opening for pumping as shown in Fig. 5.1.

Figure 5.2 is an example of detailed erosion and deposition profiles of CFC tiles used in the W-shaped divertor of JT-60U [1]. Because plasma-facing materials (PFM) are subjected to boundary plasma, erosion and deposition simultaneously occur. Hence, net erosion and gross erosion can be distinguished. In this particular case, deposition on the inner divertor is larger than the erosion at the outer divertor. However, the mass balance between the erosion of the outer divertor and deposition at the inner divertor was missed with excess deposition at the inner divertor, which was attributed to the erosion of the first wall.

Erosion/deposition on PFS (first wall and divertor) are heterogeneous in both poloidal and toroidal directions. Without making a detailed investigation of erosion/deposition profiles on full PFS in tokamaks, the unbalance between the erosion and deposition is hard to understand.

T. Tanabe, *Plasma-Material Interactions in a Controlled Fusion Reactor*, Springer Series in Plasma Science and Technology, https://doi.org/10.1007/978-981-16-0328-0_5

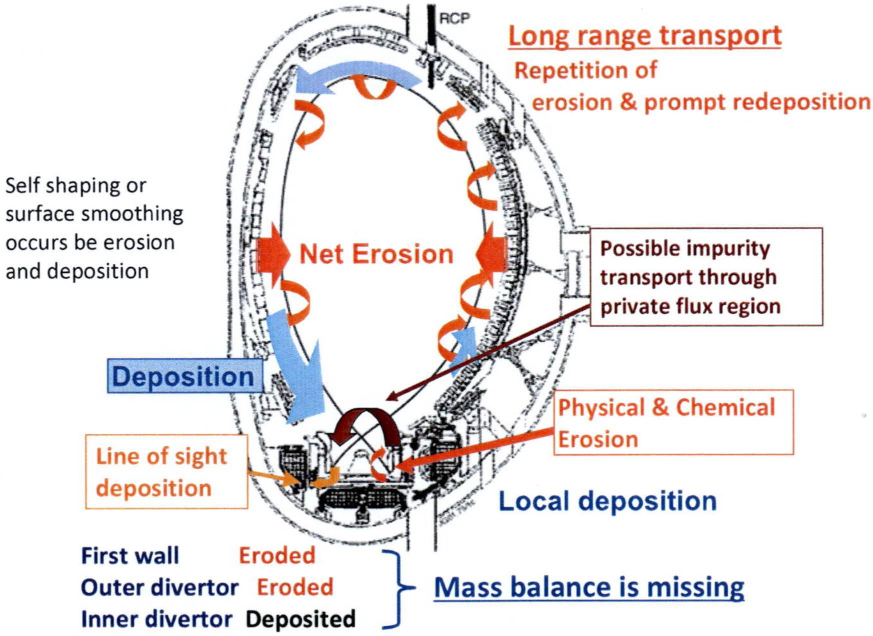

Fig. 5.1 Material transport caused by erosion and deposition in tokamak

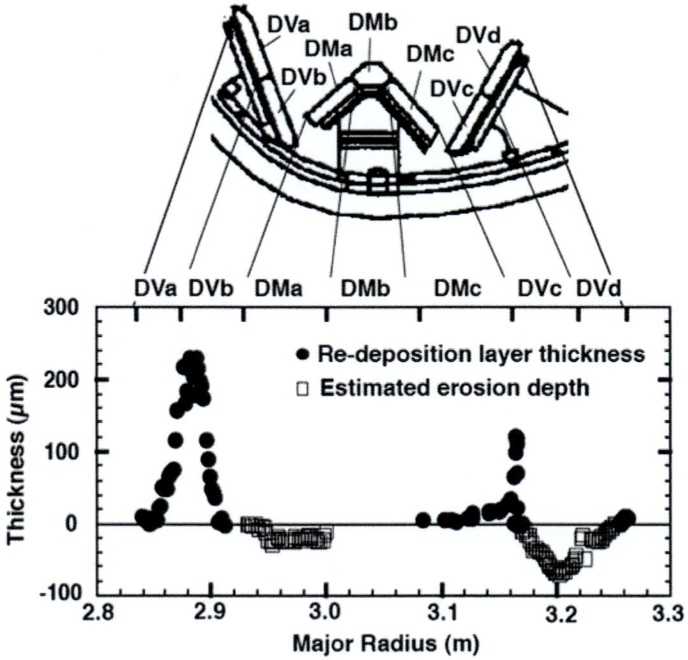

Fig. 5.2 Erosion and deposition profiles on W-shaped divertor tiles of JT-60U [1]

This chapter focuses on erosion, deposition, and transport of eroded materials in tokamaks with separation of carbon and metallic walls.

5.2 Erosion, Transport, and Deposition

Material erosion of PFS is caused by either physical sputtering, chemical sputtering, radiation-enhanced sublimation, thermal sublimation, or simply evaporation. The erosion rate changes significantly depending on masses of both incident particles and PFM, and PFM temperature. Depending on its mechanism, sputter erosion is divided into physical and chemistry sputtering. In the former, sputtered particles have energy higher than a few eV given by the incident particles and secondary particles produced by multi-collisional processes. While the energy of chemically sputtered particles is lower than a few eV or their temperature is near to the surface temperature.

In general, sputtering by hydrogen is small and most of the physical sputtering is caused by impurity ions in plasma, like C and O. Sputtering by seeded ions such as N, Ne, and Ar introduced for edge cooling is concerned. In addition, once PFM are eroded, the eroded materials penetrate in boundary plasma and contribute to sputtering of PFM (self-sputtering), which was concerned with the utilization of high Z materials as PFM. For C and Be used as PFM, chemical sputtering results in significant erosion of PFM which could limit their lifetimes.

Energies and species of eroded materials (either by physical or chemical sputtering, radiation-enhanced sublimation, or evaporation) are distributed in wide ranges. However, except those released with very high energy and penetrating deep into the plasma, most of the released particles are neutrals and they are subsequently ionized in the boundary plasma. After the ionization, owing to the strong magnetic field in a reactor, they gyrate along magnetic field lines and mostly returned to PFS as schematically shown in Fig. 5.3a. Since the gyro-radius of heavier ions is larger than that of lighter ones, the heavier ions are easier to be deposited resulting in less net erosion. Figure 5.3b shows trajectories of 50 typical sputtered Mo in divertor computed by WBC code; for particles launched near the middle of ALCATOR C-MOD outer vertical divertor made of Mo, for 800 kA shot in ohmic (OH) phase [2]. As seen in the figure, since the angle of the magnetic field line to the plasma-facing surface is very shallow, most of the ions return to the surface to be deposited. This is referred to as "prompt deposition". With the increasing ionization state from Mo^{+} to Mo^{3+}, the gyro-radius becomes smaller and transported longer, as seen in Fig. 5.3b. In addition, because sputtering of heavier or higher Z materials is less than that of lighter materials, their total erosion is much less than that of lighter or lower Z materials.

Once deposited somewhere on PFS, the deposits are subjected to plasma again. Depending on the locations of the deposits, either eroded area exposed higher incident particle flux or deposited area exposed less particle flux, they are re-eroded

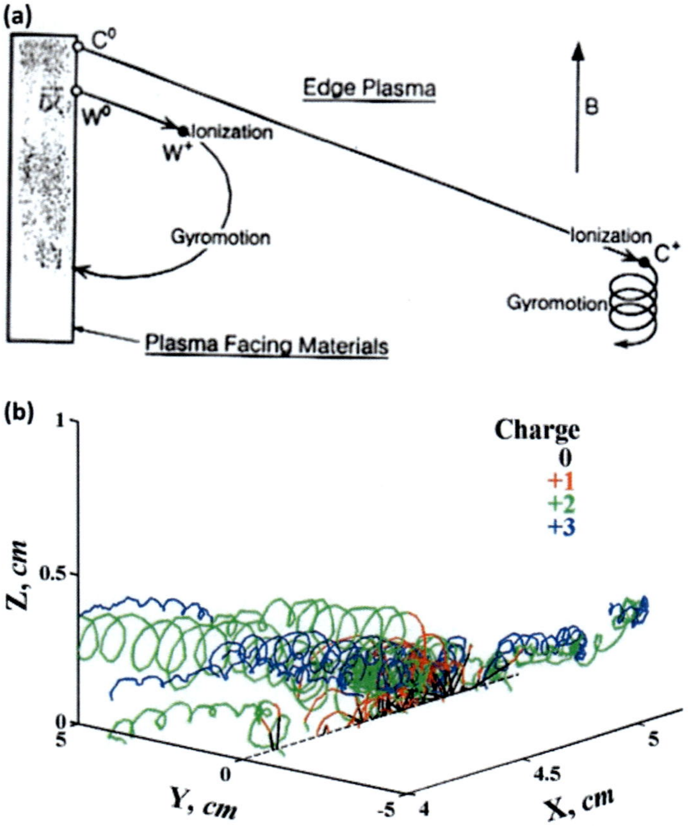

Fig. 5.3 **a** Schematic drawing showing prompt deposition with the comparison of carbon (C) and tungsten (W). **b** Trajectories of 50 typical sputtered MO in divertor computed by WBC code; for particles launched near the middle of ALCATOR C-MOD outer vertical divertor, for 800 kA shot OH phase. Coordinates: "X" along divertor from top to bottom (X = 4.5 cm is 50.5 cm poloidally below midplane, at tile 17), "Y" along the toroidal magnetic field, and "Z" perpendicular to divertor (note scale differences). (Trajectories computed for full divertor zone; $0 \leq X \leq 11$ cm, $-1.9 \leq Y \leq 1.9$ m, $0 \leq Z \leq 5$ cm; but shown here for the partial region only) (reprinted with permission from [2])

or continuously deposited, respectively. Repeating erosion and deposition, and re-erosion and re-deposition, eroded materials are transported to a longer distance, and final deposition and erosion profiles become as schematically given in Fig. 5.1 and as observed in the divertor region of JT-60U in Fig. 5.2.

There is another way for the transport of eroded materials. In boundary plasma, there is a flow of neutral fuel particles or residual fuel gas toward pumping ducts. Eroded particles escaping from the prompt deposition in the boundary plasma can be transported to plasma shadowed area or remote area following this flow to make deposits. A typical example of this transport appears as carbon transport from the outer divertor to the inner divertor through the private flux region in JT-60U with

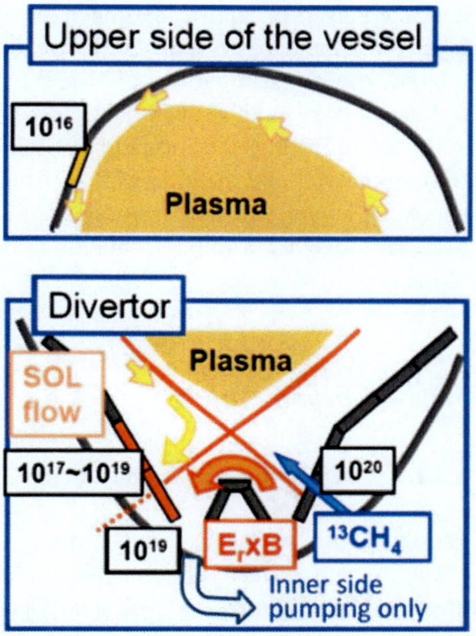

Fig. 5.4 Schematic view of the ^{13}C transport in the $^{13}CH_4$ gas puffing experiment in JT-60U. Values indicated in the squares show the order of ^{13}C surface density on each sampled tile. **a** Upper side of the vessel and **b** the divertor region (reprinted with permission from [3]). See also (Fig. 5.17)

inner side pumping as shown in Fig. 5.4 [3]. Even for W used as the outer divertor as a maker in JT-60U, W transport from the outer divertor to the inner divertor through the private flux region was observed which is described in Sect. 5.3.2 (see Fig. 5.17) [4]. This kind of direct transport through private flux region and boundary plasma is quite dependent on the geometrical structure, and different deposited patterns were observed in different plasma apparatus as shown later.

5.3 Formation of Deposited Layers Made of Eroded Materials

The formation of deposited layers and their microstructure depend on the energy and direction of particles injecting into PFS and the temperature of PFM. Moreover, the difference in chemical affinity to hydrogen between carbon and metals results in significant differences in the microstructure of the deposited layers and their hydrogen retention. In the case of C, the deposited layers retain large amounts of D and T, which decrease with temperature from 0.4 in an atomic ratio of (D+T)/C below 500 K to less than 0.01 above 1000 K. Accordingly, the structure of the deposited layers changes from amorphous hydrocarbon like below 500 K to graphite like above 1000 K. The former becomes soft layers with less thermal conductivity, while the latter hard layers with higher thermal conductivity enhancing graphitization of the deposits. It is noted that the content of O in the C deposits is small owing to the release of CO. For metallic deposited layers except Be, hydrogen retention is much less than that in C-deposited layers, and they are crystallized. However, metallic deposited layers often contain a large amount of impurities of O and C and show amorphous-like structure that contains a large amount of D+T because O and C in metals trap hydrogen.

In the following, the formation of the deposited layers is described separately for carbon wall and metallic wall focusing basic process for the formation of the deposited layers. In Chap. 8 is revisited erosion and deposition observed in large tokamaks together with the formation of dust mostly resulting from the exfoliation of the deposited layers.

5.3.1 *Carbon Wall*

5.3.1.1 Deposition on Plasma-Facing Surface

Erosion occurs at plasma-facing surface, while deposition at various places such as plasma-facing surface, area on the line of sight from plasma, plasma shadowed area, and remote area from plasma. Detailed observation of deposited layers in Fig. 5.5 gives the information on how deposited layers were formed in sequential 40 discharges in JT-60 in which most of the plasma-facing surface was covered by graphite tiles [5]. The deposited layers were mostly made of carbon and showed layered structure with each layer corresponding to a discharge indicated in the left. Compared to limiter discharges, NBI heated divertor discharges gave thicker layers showing columnar structure because of their higher power load resulting in higher erosion and deposition and temperature rise. Even disruption added a layer including small black dots probably corresponding to particles or dust exfoliated from PFS.

Deposited layers retain a significant amount of hydrogen because during deposition hydrogen is incorporated in them and after deposition, the layers are exposed to boundary plasma and residual fuel gas. Since the temperature of plasma shadowed area and remote area is lower than that of PFS, hydrogen concentration in the deposited layers at these areas is much larger than that in the deposited layers on PFS and eroded area.

On JET divertor CFC tiles used in the DTE campaign with DT discharges, deposited profiles in divertor area well corresponded to T retention profiles as seen in Fig. 5.6. The figure shows the T profiles and the photos of JET divertor tiles (numbered from BN 1 to BN 10) [6, 7]. The T profiles were quite consistent with the C-deposited profiles seen in the photos. The numbers in the divertor structure in the center are the amount of T in mg retained in each CFC tile. The highest retention was observed in the deposited layers on the louvers of the pumping duct which were not directly exposed to plasma but exposed to neutral or gas flow pumped out. The high T retention also appeared at the bottom of the BN 4 tile which was shadowed from plasma by BN 3 and heavily deposited. Inhomogeneity in the T profile on the shadowed area was caused by the exfoliation of the deposited layers very similar to the T profile of TFTR tile in Fig. 5.7 [8]. The lowest T retention appeared at the

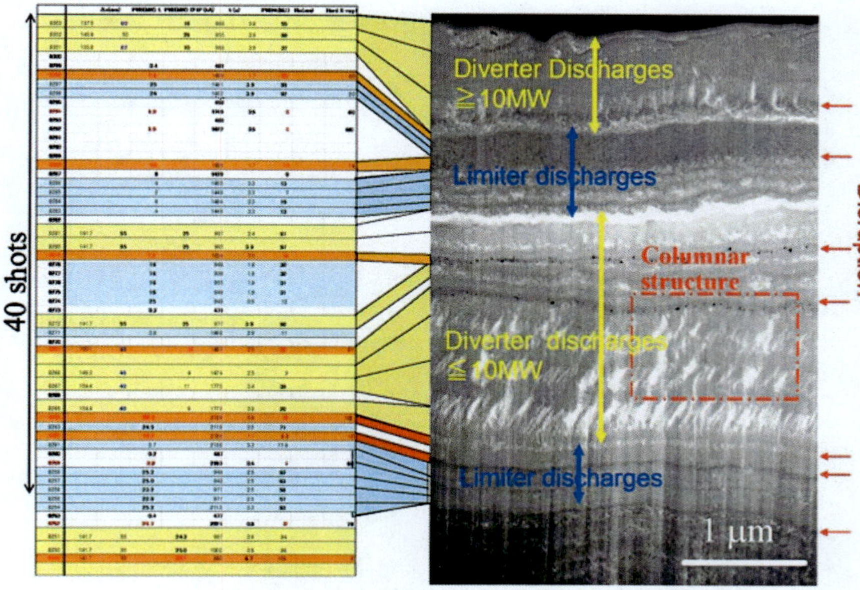

Fig. 5.5 TEM observation of deposited layers on JT-60 divertor produced by around 50 discharges with a comparison of discharge history [5]. Each discharge made a deposited layer of which thickness is different depending on discharge modes with thicker ones for divertor discharges and thinner ones for limiter discharges. Lower power divertor discharges made columnar structured layers remain in the incoming direction of deposited materials, while higher power discharges changed the columnar layers to planer structure owing to the annealing effect of their higher power load. See also (Fig. 5.8)

divertor target area in BN5, which was eroded. It should be noted that in the toroidal direction, T retention and C deposition were also quite inhomogeneous. Because of tile alignment to avoid edge effects as shown in the inset (bottom left), the right side of the inner divertor at the high field side was erosion dominated (less T), while the left side was deposition dominated. The profile is opposite on the outer divertor tiles at the lower field side.

D-T discharges in TFTR also remained that T profiles are strongly correlated to carbon deposition profiles as shown in Fig. 5.7 [8]. Since TFTR is a limiter tokamak, eroded area and deposited area on the inner surface or bumper limiter are clearly distinguished. Some part of the deposits was exfoliated showing little T retention beneath the deposits. The sides of the eroded tile show higher T retention so as the higher C deposition. Since the tile sides were facing tile gaps, T retention in tile gaps in a reactor becomes a concern. However, different from TFTR, deposition in tile gaps are not so significant in divertor tokamaks like JET and JT-60U as TFTR. The T retention in the gap is discussed later. It is important to note that the T profile of the eroded tile reflects characteristics of a CFC tile composed of fivers and matrix (compare the photograph and the T profile of the eroded tile in the figure). That is because both the amount of retained T and erosion were different between the fiver and the matrix.

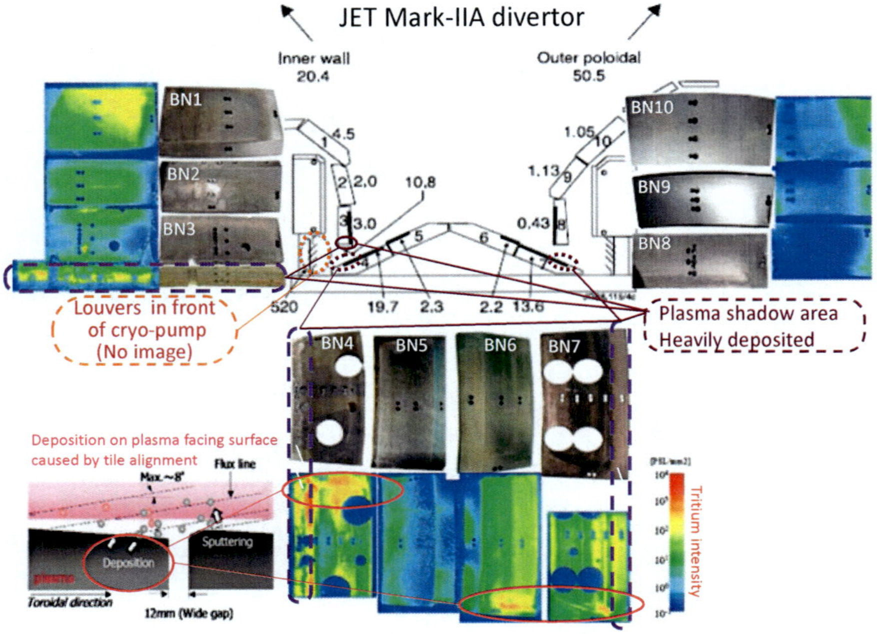

Fig. 5.6 Tritium (T) distribution of floor tiles of JET mark IIA divertor used in DTE-1 campaign, and the numbers in the divertor figure are the amount of T (in mg unit) retained in the drilled column in each tile. A significant amount of T was retained in deposits on plasma shadowed area, in particular, louvers in front of the pumping duct, the side of tile BN3 facing wide opening, and fringes of tiles BN4 and BN7. Steps between neighboring tiles to avoid edge heating gives toroidal asymmetry in deposition profiles on the base tiles as shown in the inset [6, 7]

5.3.1.2 Modification of Deposited Materials

In the earlier investigation of deposits (deposited materials) in tokamak plasma apparatus, it was shown that they consisted of various elements included in materials used as its vacuum vessel and impurities, such as C, O, K, Na, Al, Cr, Fe, Ni, W, and many others and were referred to as "tokamikium" [9]. After selecting carbon as PFM, the deposits mostly consisted of carbon and hydrogen. Still some constituent elements of the vacuum vessel, like Fe and Cr, are included. Since the deposits were overlaid with shot by shot, they often showed layered structure as demonstrated in Fig. 5.5 and referred to as deposits or redeposited layers. However, temperature rise caused by high-power load modified their structure. Figure 5.8 demonstrates structure modification of the deposited layers on JT-60U divertor tiles [1]. Different from Fig. 5.5, which clearly shows layer by layer deposition with sequential discharges, the structure shows a strong influence of temperature rise after the deposition. In this particular case, because of the porous nature of the deposited layers, thermal contact of deposited layers made at earlier discharges to the layers made afterward was so poor to make the temperature of the latter layers very high, and consequently, they

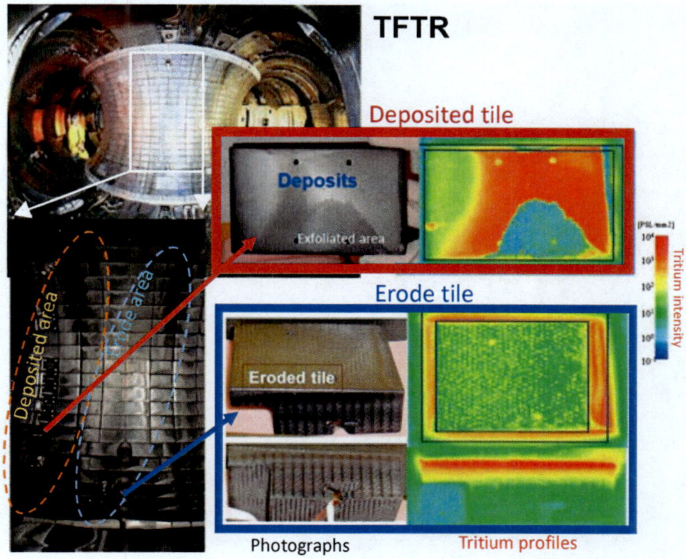

Fig. 5.7 Tritium profiles on eroded and deposited tiles used as bumper limiters in TFTR [8]. In the deposited tile, some deposited layers were exfoliated which clearly indicated that T was mostly in the deposited layers. On the eroded tile, T retention on the plasma-facing surface was quite small, while its tile sides were deposited and an appreciable amount of T was retained. Higher erosion on the tile surface gave a higher deposition on its tile sides. It is noted that the T profile of the eroded tile shows CFC structure because T retention was different between fiber and matrix in CFC

were transferred to different ordered structure reflecting nature of layered structurer of graphite. Once such ordered structure was made, the thermal conduction along the parallel direction of the layers became much better than the normal direction to porous layers and enhanced the ordering within the layers.

The microstructure of the deposited carbon (C) layers was very complex because C takes polymorphism due to three different bonding forms, sp, sp^2, and sp^3. The inclusion of hydrogen also modifies their structure. In large tokamaks, such as JT-60U and JET, the structure of deposits is significantly different depending on the location, i.e. directions of the coming particles to be deposited and how high the temperature they were subjected to. The ordered structure that appeared in JT-60U (Fig. 5.8) was graphitized and less H retention owing to temperature rise, sometimes over 1300 K. In addition, depending on where carbon sources were or where erosion occurred and how they were transferred, the structure of the deposited materials was modified. As an example, the columnar structure in Fig. 5.8 was caused by the deposition with the incident angle of 60°.

Cyclic heat load and transient heat load give thermal stress between the deposited layers and the substrate. Correspondingly, some area of the deposited layers is exfoliated as flakes as seen in Fig. 5.7. The flakes could turn to dust often observed as small white dots appearing during plasma discharges [10–12]. Examples of the dust collected from JET-ITER-like wall are given in Fig. 5.9 [12]. Not only the elements

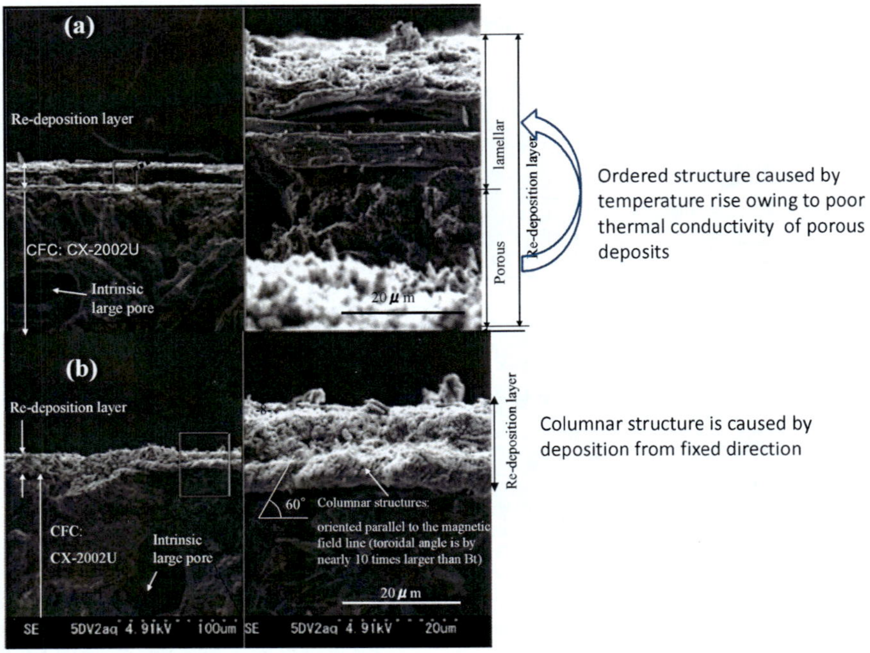

Fig. 5.8 Changes of microstructures of deposited layers on CFC tiles (CX-200U) with incident flux, angle of incidence, and temperature [1]. The columnar structure that appeared in (**b**) at lower power divertor discharges remained in the direction of incident carbons to be deposited, while ordered structure in (**a**) was caused by the restructuring of porous deposits to graphite-like layered structure with temperature rise owing to their poor thermal conductivity

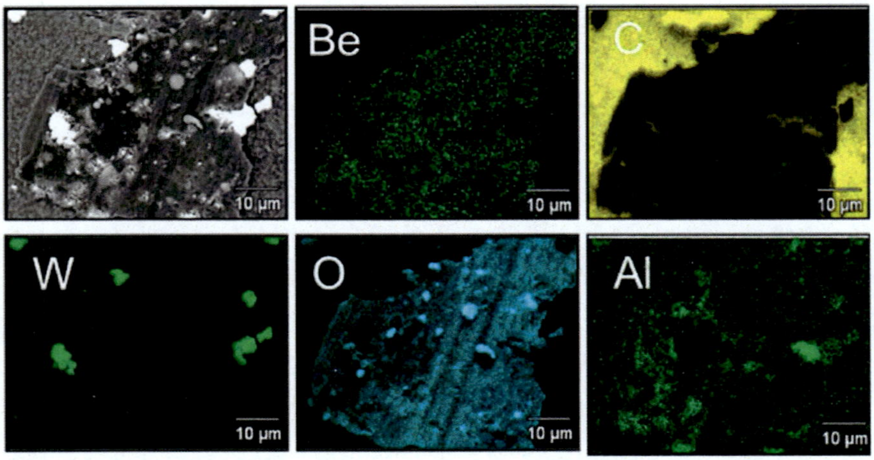

Fig. 5.9 Beryllium flakes detached from the deposit and the distribution of respective elements in the studied area in JET-ITER-like wall (reprinted with permission from [12])

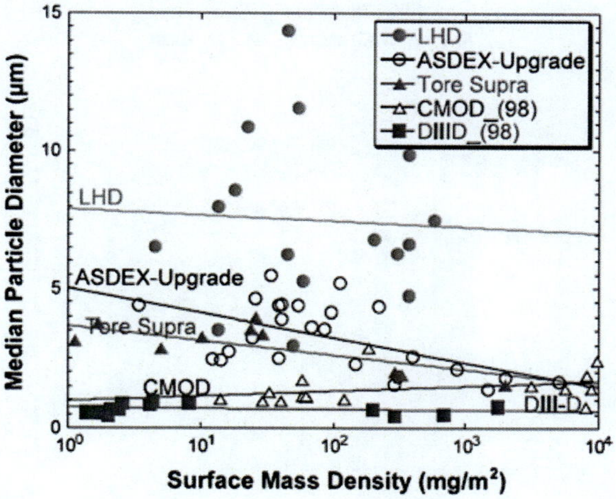

Fig. 5.10 Comparison of dust sizes versus surface mass density for various fusion devices. A linear fit is provided from the data to aid in viewing the relative sizes. LHD and ASDEX-Upgrade have comparatively large scatter in the size data (reprinted with permission from [13])

used as ITER-like wall, W and Be, but also remaining elements in the JET vessel, C, O, and Al, were included in the dusts. Collected dusts in various plasma machines are compared in terms of their sizes and surface mass densities in Fig. 5.10 [13]. Probably because of different origins of the dusts, exfoliation of deposited layers, droplets, surface cracking, and others, their sizes and constituents were widely distributed. Since dusts in a fusion reactor include T and neutron-activated materials, they must be very hazardous and safety concern. The formation of dusts and their influence on plasma are discussed in Chap. 8 (Sect. 8.3) in detail.

5.3.1.3 Deposition on Non-plasma-Facing Surfaces

As already described, eroded atoms and molecules are immediately ionized, gyrated, and transported along magnetic field lines in boundary plasma. Then they are injected to plasma-facing or non-plasma-facing surfaces to be deposited layers. Except for those area exposed to plasma particle flux high enough to re-erode the deposited layers, the deposited layers are piling-up, making a clear separation between net deposited areas and net eroded areas on PFS. In addition, gyrated ions can penetrate into tile gaps and get deposited at tile side surfaces facing the gap and the bottom of the gap [14]. Detail of deposition profiles on the sides of eroded tile given in Fig. 5.7 is shown as the T profiles in Fig. 5.11. The depth profiles along lines from the entrance to the bottom of the gap are given in the right. Because the T profiles in TFTR well corresponded to the C deposition profiles, the T profiles along the tile gap or tile sides represent the C deposition profiles. The deposition profiles decayed from the front surface to the bottom of the tile gap showing two exponential decays that are nearly the same on all tile sides. The first decay profile was well simulated with the repetitive processes of erosion and prompt deposition of C initially eroded at the front surface, while the second one with a longer decay was caused by deposition from plasma

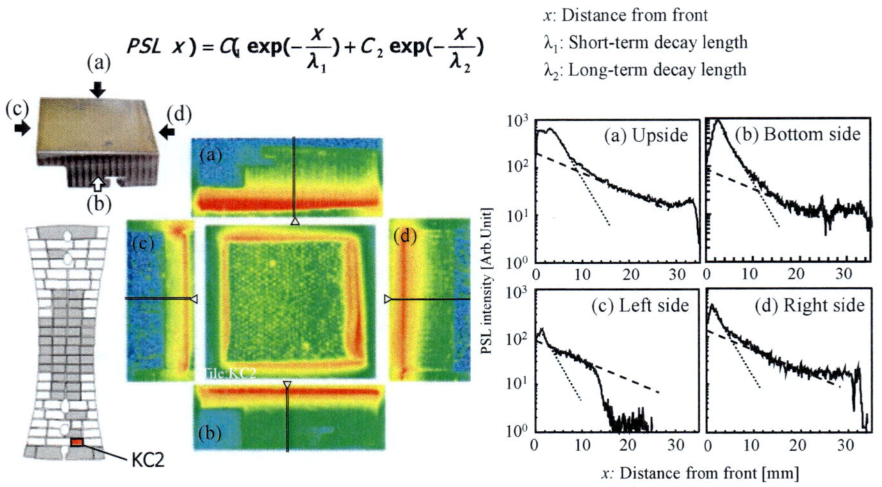

Fig. 5.11 T profiles on gap-facing sides of the eroded tile in TFTR is given in Fig. 5.7, which was quite consistent with the C deposition profiles [8]. Most of the deposits appear near the entrance of the gap caused by prompt deposition of eroded C at the front surface showing exponentially decays with the gap depth. In addition, penetration of boundary plasma gave deposition with longer decay profiles [14]

penetrated in the gap [14]. The decay length of the shorter one changed with not only the width and depth of the gap but also the location of the gap. Figures 5.12 and 5.13 show C deposition profiles determined for T profiles in tile sides facing toroidal side and poloidal side, respectively, in JT-60U [15]. The deposition profiles on the toroidal sides (Fig. 5.12) were dominated with the repetitive processes of erosion and deposition with the decay length of about 3 mm irrespective of the location, while the tile sides facing larger opening (b) and (c) in Fig. 5.13 were dominated by plasma deposition or transport of neutrals without appreciable decay. This is quite similar to the significant deposition on louvers and tile sides facing a large opening in JET divertor (see Fig. 5.6). In boundary plasma, there are neutral flows to pumping ducts. Ionized ions and molecules are neutralized by charge exchange or recombination with electrons to be neutralized. Then, they are transported along the neutral flow to the plasma shadowed area and remote area. In the case of C-wall, eroded C and hydrocarbons are transported along this flow and consequently deposited at the plasma shadowed area but on the line of sight from plasma.

This kind of line-of-sight deposition was also observed on the bottom of the tile gaps. Figure 5.14 shows outboard first wall with and without armor tiles in JT-60U. The deposition patterns were different with the location of the armer tiles because the bottom of the tile gap was closed and hence penetrating plasma into the tile gaps was different depending on the location of the gap and the width and depth of the gap. Furthermore, at the bottom of the gap, neutral H pressure changes with the gap width and depth (see the inset in Fig. 5.14). Because the bottom of the gap was closed, the penetrating plasma particles are neutralized and gas pressure at the

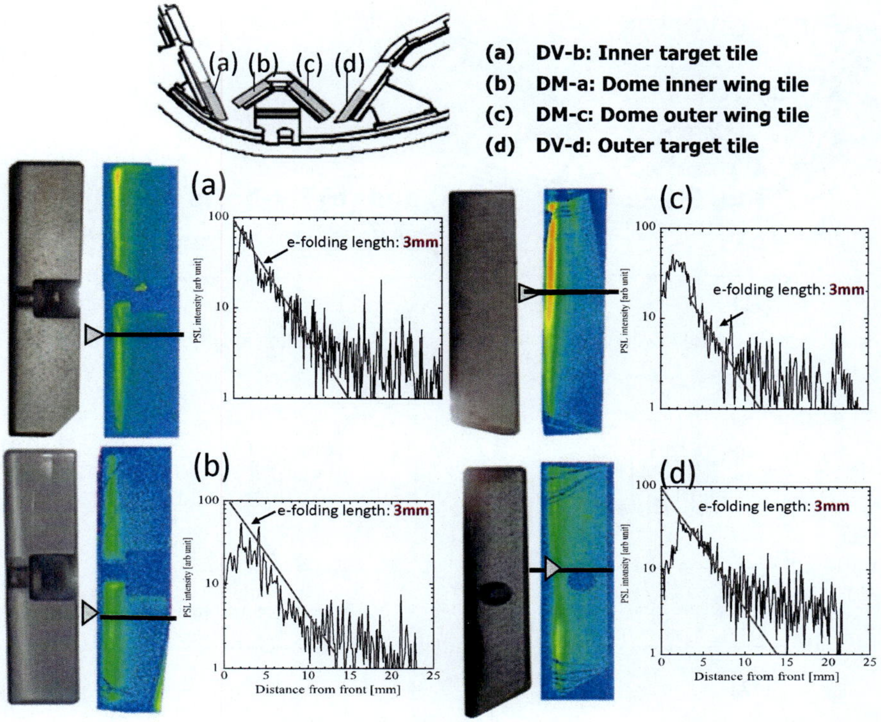

Fig. 5.12 Photo images and T profiles of toroidal sides of divertor tiles of JT-60U. Thickness profiles along lines from the entrance to the bottom of the gap show the same exponential decay of 3 mm caused by prompt deposition of C eroded at the front surface

bottom could be higher than that at the entrance. Since chemical sputtering of C by H has no threshold energy, higher hydrogen pressure could re-erode once deposited C. Accordingly, the bottom of the narrow gap tended to be eroded and the wider gap deposited corresponding to possible higher pressure in the latter. It should also be noted that deposition appeared even behind the tile, i.e. at the interspace between the tile backside and SS base, which were too narrow for escaping re-erode hydrocarbons. Thus, deposition profiles in the shadowed area and remote area are quite dependent on geometries of the areas so that it is hard to describe the general picture of carbon transport (erosion and deposition) in tokamaks.

In the remote area, the line-of-sight deposition was clearly observed behind the divertor in JT-60U as shown in Fig. 5.15 in which photographs of materials probes are set at the respective location behind the W-shaped divertor and exposed to plasma [15]. Only the probes of ①–④ which were set at the line of sight from the plasma show C deposition and the others remained shiny. Thus, C deposition at the remote area is limited to those areas on the line of sight from the plasma and their further transport is prohibited because of no or very thin plasma there. Although JT-60U was operated with PFM around 420 K, the temperature of non-plasma-facing surfaces was

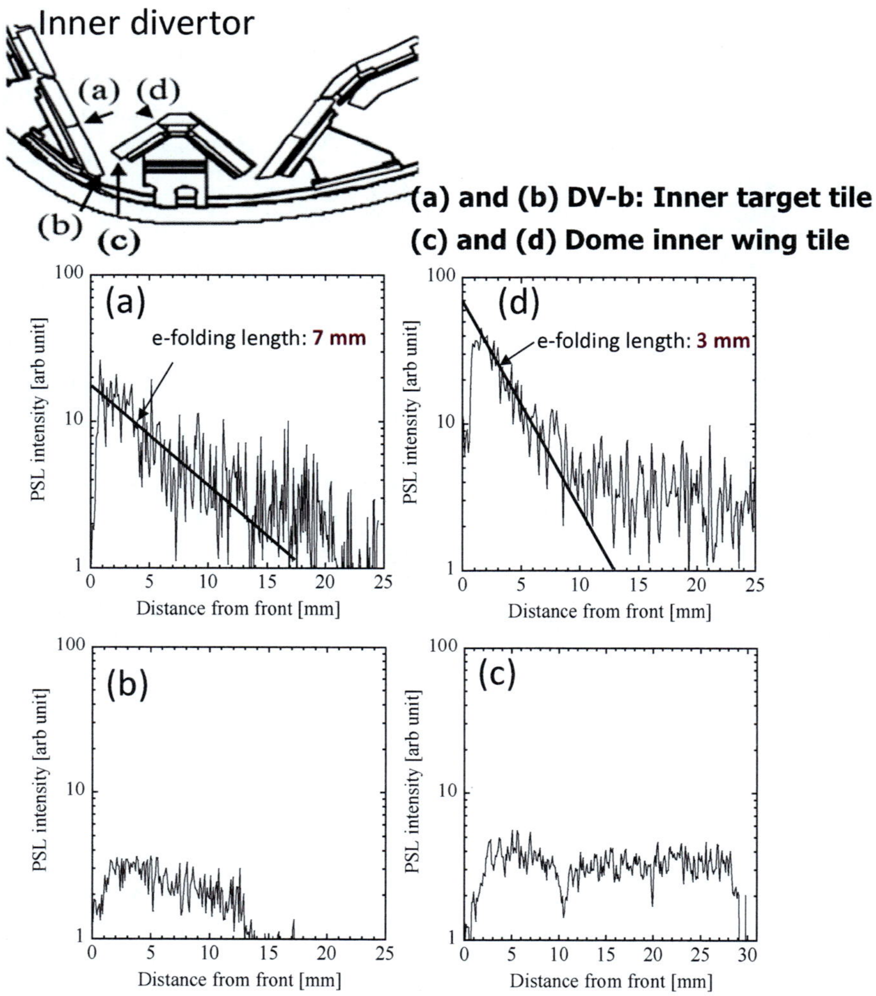

Fig. 5.13 Deposition profiles obtained from T profiles on poloidal sides of the inner divertor tiles of JT-60U. The profiles are different for sides facing the tile gap and those to the wider opening, with the former showing the exponential decay, while the latter rather uniform deposition. (See the text)

lower, and accordingly, the deposits there retained a higher hydrogen concentration of around 0.6–0.8 in H+D/C atomic ratio, which made the structure of the deposit to be amorphous and also succeeding deposition easier. Thus, deposition on the remote area was different from that of the tile gaps and/or plasma shadowed area but on the line of sight from plasma.

The deposition given by the neutral flow is more clearly seen in the outer dome wing tile of the W-shaped JT-60U divertor with inside pumping only as shown in Fig. 5.16 [1]. The figure indicates that the eroded materials are released from the

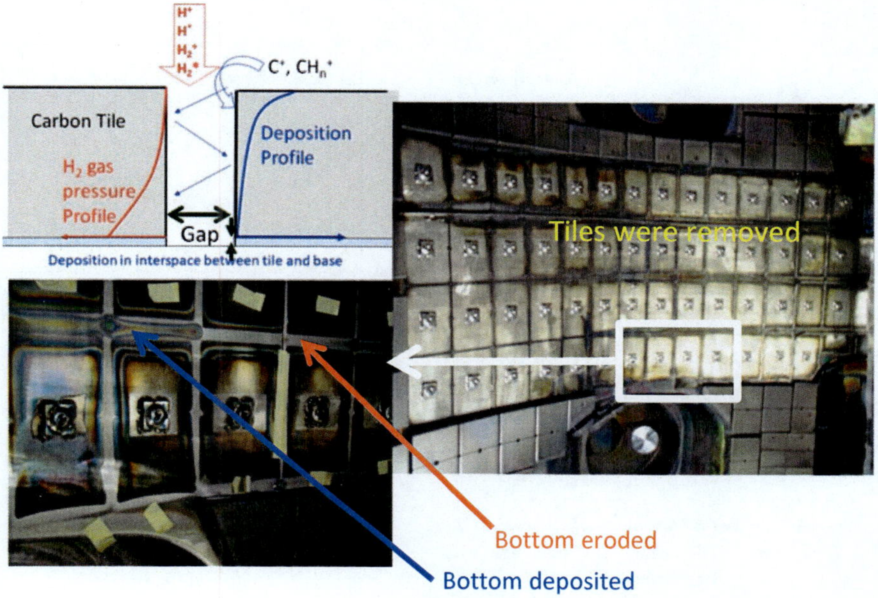

Fig. 5.14 Erosion and deposition observed at the bottom of the tile gaps and interspace between the tiles and the base of the outboard first wall in JT-60U. The inset schematically shows deposition profiles on the tile sides and possible pressure change of H_2 in the gap

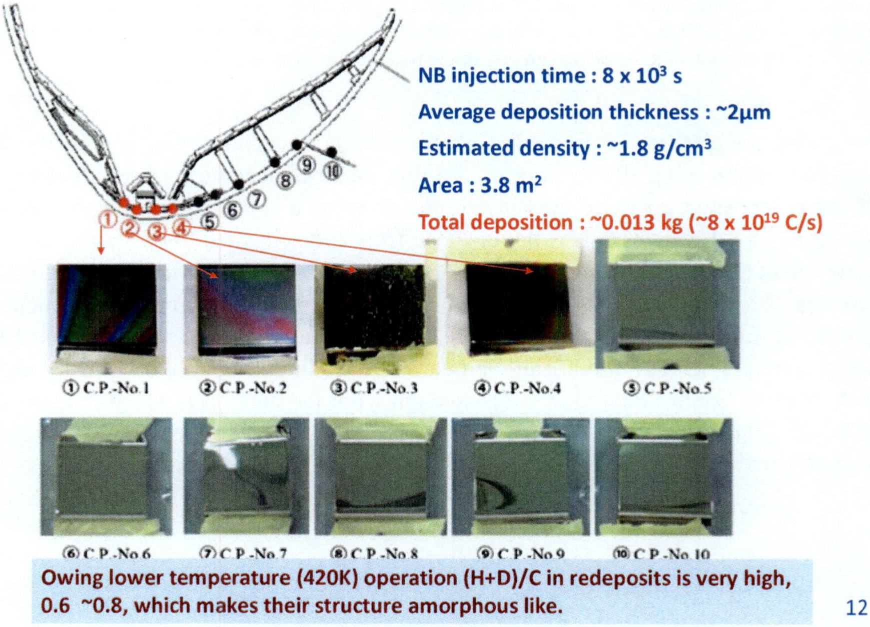

Fig. 5.15 Line-of-sight deposition observed on probe samples set behind the divertor in JT-60U. [15]

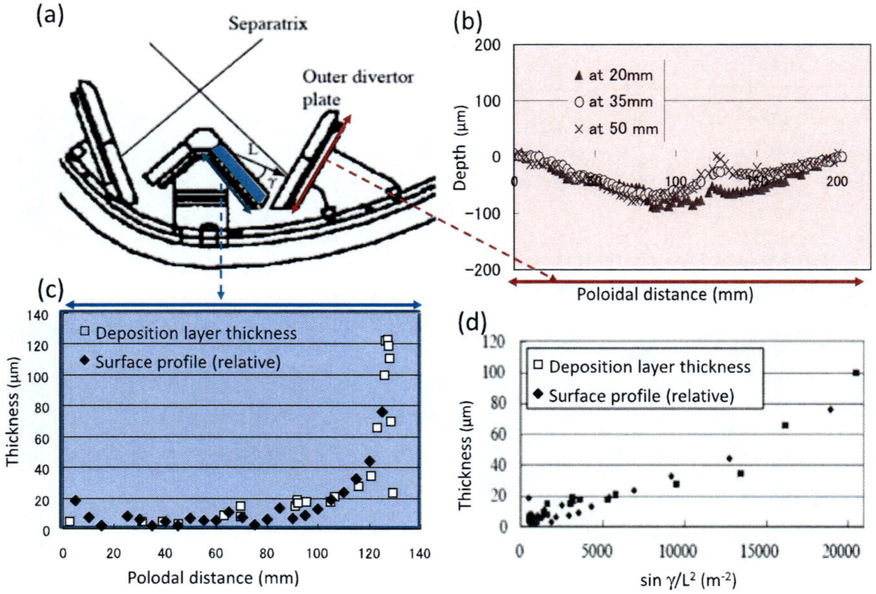

Fig. 5.16 Direct transport of eroded carbon from the outer divertor to the outer dome wing to be deposited. **a** Geometry of the W-shaped divertor of JT-60U, **b** Erosion profile on the outer divertor, **c** Deposition profile on the outer dome wing tile, and **d** Dependence of the thickness of the redeposited layers of the outer dome wing on a solid angle from the outer divertor (sin γ/L^2), where L is the distance from each eroded location on the outer divertor tile to the deposited point on the dome wing. The linear dependence indicates that eroded materials are released in 2π direction following cosine low and deposited directly to the outer dome wing tile [1]

outer divertor plate in 2π direction following cosine low and deposited directly to the outer dome wing tile as appeared as the linear dependence of the thickness of the redeposited layers of the outer dome wing on a solid angle from the outer divertor (sin γ/L2), where L is the distance from the deposited point on the dome wing tile to each surface location of the outer divertor tile. The structure of deposits (see Fig. 5.17c) indicates the direction of incoming carbon or hydrocarbon particles (declining from the normal direction) to be deposited which are originated at the outer divertor tile. The flow of eroded carbon over the private flux region (dome wing tiles) is confirmed with $^{13}CH_4$ injection from the outer divertor as discussed in Fig. 5.4. Significant deposition caused by the neutral flow to the pumping duct is also observed on louvers of JET divertor as very high T retention in Fig. 5.6.

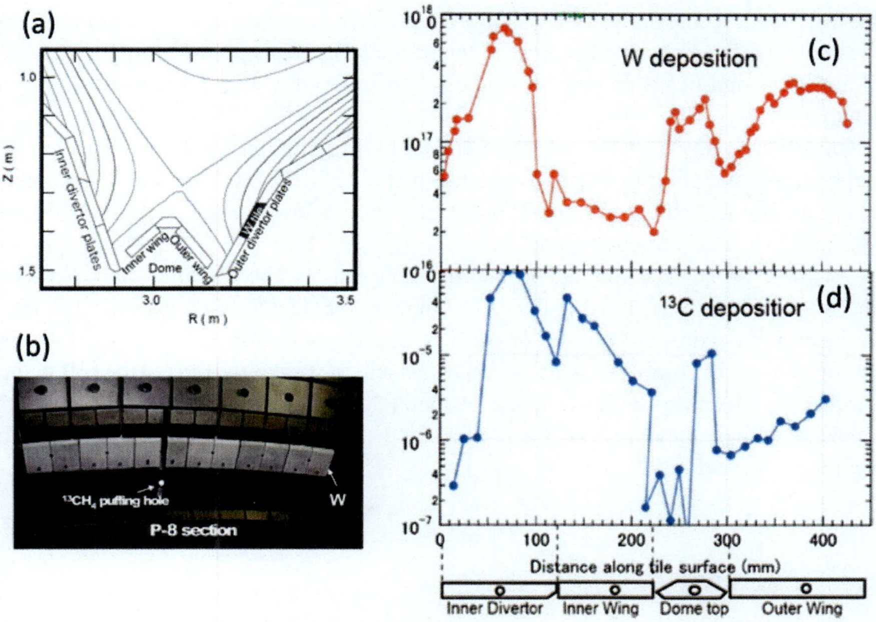

Fig. 5.17 W transport through private flux region in the W-shaped divertor in JT-60U. **a** Geometry of JT-60U divertor. W-coated tiles were installed in the outer divertor at the P8 toroidal section, where $^{13}CH_4$ gas puff was made separately (see Fig. 5.4) as shown in (**b**). **c** and **d** are poloidal distributions of deposited W and ^{13}C, respectively on the divertor tiles (reprinted with permission from [4])

5.3.2 *Metallic Wall*

In case of W-wall, owing to the lack of H chemical sputtering, its erosion would be dominated with physical sputtering by impurity ions and seeded gas ions introduced for cooling of the boundary plasma. In addition, the large gyro-radius of W after the ionization enhances prompt deposition nearby the eroded location at the plasma-facing surface. Although self-sputtering of W is used to be concerned, recent experiments using high Z walls have shown that low erosion of W does not make W concentration in boundary plasma high enough to induce a self-sputtering cascade. Consequently, the deposition of W is much less compared to C. Nevertheless, W transport from the outer divertor to the inner divertor through the private flux region was clearly observed in JT-60U with using W coating on one of the outer divertor tiles in Fig. 5.17 [4]. As seen Fig. 5.17c, W deposition on the outer dome wing tile and the inner divertor tile was appreciable and its deposited profile was similar to that of ^{13}C given in Fig. 5.17d which was caused by direct transport through the private flux region of gas puffed $^{13}CH_4$ at the P8 port (W-coated tiles were at the same port) as indicated in Fig. 5.4. Different from C, owing to the very low vapor pressure of W, long-range transport of W carried by neutral flow though boundary plasma was hardly seen. Accordingly, W deposition at plasma shadowed area and remote area was not significant. It should be noted that if the W deposits included impurities like C and O, a significant amount of H should be retained as discussed in Sect. 9.4 in Chap. 9.

In the case of Be, owing to chemical sputtering making BeH_2, its erosion and deposition are similar to those of carbon. In JET ILW (ITER-Like Wall), Be dominates in the deposits, and deposition at plasma shadowed area is appreciable, of which detail is described in Chap. 8.

Erosion by surface melting is another concern for the metallic wall. Surface melting layers move following the electromagnetic field and often leave as droplets. Adhesion of the droplets on PFS or surfaces of components in a reactor is not strong enough so that the droplets easily become dust and move around during the following discharges. Because the dusts made of metals contain neutron-activated elements together with T, they are much hazardous and concerned with safety compared to dust made of carbon.

5.4 Summary

Erosion and deposition behavior/mechanism is quite different between PFMs, hydride-making, and non-hydride-making elements. For carbon wall, chemical erosion through hydrocarbon formation results in a significant amount of erosion which shortens the lifetime of the C-wall and its deposits retain a large amount of hydrogen. On the other hand, small erosion of W appreciably reduces deposits and small hydrogen solubility in W also reduces hydrogen retention. That is the main reason for the selection of W as divertor tiles of ITER.

The above conclusion is based on the data obtained in present tokamaks which are operated at RT or ambient temperature much below the operation temperature of a reactor. Since high temperature use of C and W would change erosion and deposition scheme and hydrogen retention as well, data accumulation under higher temperature are awaited for the selection of PFM (see Chap. 10).

Although the deposits retain a large amount of hydrogen to be the main cause of T inventory in a reactor, higher temperature reactor operation could change the retention scheme which is discussed separately in Chap. 8 considering temperature effects.

Owing to the prompt deposition, sides of plasma-facing armor tiles or facing to gaps of the tiles are mostly deposited. In addition, eroded materials from C-wall (mostly hydrocarbons) travel a long distance to be deposited at a far remote area like pumping duct or even penetrating into pumps but mostly in the line of sight from the plasma. In this respect, installation of collector plates heatable to very high temperature at possible deposition locations on plasma shadowed or remote areas could be one of the important methods to reduce T inventory.

References

1. Y. Gotoh, T. Tanabe, Y. Ishimoto et al., Long-term erosion and re-deposition of carbon in the divertor region of JT-60U. J. Nucl. Mater. **357**, 138–146 (2006)
2. J.N. Brooks, J.P. Allain, D. Whyte et al., Analysis of C-MOD molybdenum divertor erosion and code/data comparison. J. Nucl. Mater. **415**, S112–S116 (2011)
3. Y. Ishimoto, Y. Gotoh, T. Arai et al., Transport studies of carbon impurities using $^{13}CH_4$ gas puffing in JT-60U, Proceedings of 12th *International Conference on Fusion Reactor Materials*, Santa Barbara, USA 4–10 Dec 2005 (2005)
4. Y. Ueda, M. Fukumoto, J. Watanabe et al., Localized tungsten deposition in divertor region in JT-60U. Nucl. Fusion **49**, 065027 (2009)
5. Y. Gotoh, T. Arai, J. Yagyu et al., Transmission electron microscopy of redeposition layers on graphite tiles used for open divertor armor of JT-60. J. Nucl. Mater. **329–333**, 840–844 (2004)
6. K. Sugiyama, K. Miyasaka, T. Tanabe et al., Tritium distribution on the surface of plasma facing carbon tiles used in JET. J. Nucl. Mater. **313–316**, 507–513 (2003)
7. J.P. Coad, N. Bekris, J.D. Elder et al., Erosion/deposition issues at JET. J. Nucl. Mater. **290–293**, 224–230 (2001)
8. T. Tanabe, K. Sugiyama, C.H. Skinner, N. Bekris, C.A. Gentile, J.P. Coad, Tritium Retention in the Gap between the Plasma-Facing Carbon Tiles Used in D-T Discharge Phase in JET and TFTR. Fusion Science and Technology **48**, 577–580 (2017)
9. R. Behrisch, M. Mayer, C. Garcia-Rosales, Composition of the plasma facing material Tokamakium. J. Nucl. Mater. **233–237**, 673–680 (1996)
10. N. Bekris, J.P. Coad, R.-D. Penzhorn et al., Characterization of flakes generated in JET after DD and DT plasma operations. J. Nucl. Mater. **337–339**, 659–663 (2005)
11. A.T. Peacock, P. Andrew, P. Cetier et al., Dust and fakes in the JET MkIIa divertor, analysis and results. J. Nucl. Mater. **266–269**, 423–428 (1999)
12. A. Baron-Wiechec, E. Fortuna-Zaleśna, J. Grzonka, First dust study in JET with the ITER-like wall: sampling, analysis and classification. Nucl. Fusion **55**(7), 113033 (2015)
13. J.P. Sharpe, D.A. Petti, H.-W. Bartels, A review of dust in fusion devices: Implications for safety and operational performance. Fusion Eng. Des. **63–64**, 153–163 (2002)
14. K. Sugiyama, T. Tanabe, K. Masaki et al., Tritium distribution measurement of the tile gap of JT-60U. J. Nucl. Mater. **367–370**, 1248–1253 (2007)
15. T. Tanabe, K. Sugiyama, C.H. Skinner, N. Bekris, C.A. Gentile, J.P. Coad, Tritium Retention in the Gap between the Plasma-Facing Carbon Tiles Used in D-T Discharge Phase in JET and TFTR. Fusion Science and Technology **48**(1), 577–580 (2017)

Chapter 6
Material Modification by High-Power Load and Its Influence on Plasma

6.1 Power Load to PFM

As noted in Chap. 3, power load to plasma-facing materials (PFM) can be divided into two, steady and transient loads. Figure 6.1 summarizes five main power loads in terms of their time durations and appearing frequencies in ITER [1]. In the figure, as transient heat loads, bursts of edge localized modes (ELM), disruptions, and vertical displacement events (VDE) are separated from steady heat load to divertor and first wall (FW).

With advances of plasma confinement, i.e. increasing of plasma temperature and density to satisfy the Lawson condition for D-T burning, heat load under steady operation becomes very large. Allowable power load to plasma-facing materials is depending on not only the materials themselves but also cooling power. In current techniques, the maximum heat removal by a water-cooling system would be around 20 $MW{\cdot}m^{-2}$ as described in Chap. 3. Therefore, reduction of power load to be less than 20 $MW{\cdot}m^{-2}$ is necessary and the power exhaust to divertor is one of the hardest tasks in ITER and reactors [2, 3].

Due to such high-power load, only limited high-temperature materials can be used as PFM or plasma facing armor tiles covering structure materials. They should have high melting point, low vapor pressure, high thermal conductivity, and heat shock resistance. Only W, Carbon-based materials, and some composite materials like SiC are PFM candidates. Still material erosion due to sputtering, sublimation and particle release due to surface cracking caused by thermal heat shock cannot be avoided.

As already discussed in Chap. 2, concerned power loads to PFM from plasma consist of radiations and particles from plasma core and edge, while the power load by neutron is not significant. However, the continuous load of 14 MeV neutrons produces various kinds of radiation damages and will degrade all materials used in a fusion reactor. This is the subject of an extensive long-term materials test program [4].

The expected power load by giant ELM is of the order of 1 $GW{\cdot}m^{-2}$ within a sub-millisecond time scale [5, 6]. Up to now, only limited information is available on the material performance under these events. VDE is transient events that may deposit

T. Tanabe, *Plasma-Material Interactions in a Controlled Fusion Reactor*, Springer Series in Plasma Science and Technology, https://doi.org/10.1007/978-981-16-0328-0_6

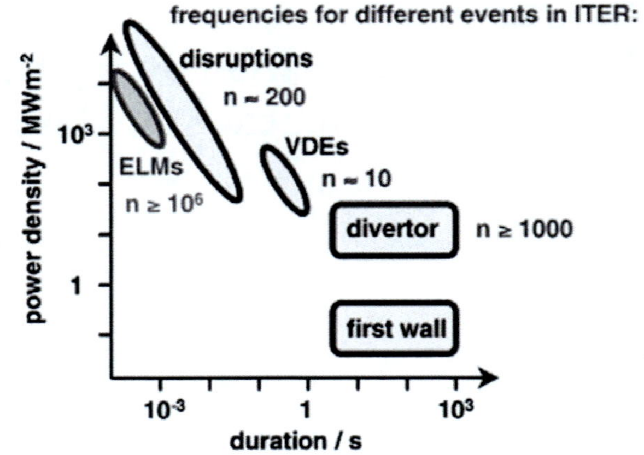

Fig. 6.1 Plasma-induced power loads on PFCs in ITER; n is the expected frequency for these events (reprinted with permission from [1]). Bursts of edge localized modes (ELM), disruptions, and vertical displacement events (VDE) are separated from steady heat load to divertor and first wall

a large fraction of the plasma energy on relatively small wall areas [7]. Disruption, which often accompanies arching [8, 9] and run-away electrons [10], is spontaneous release of confined energy in plasma to limited area within millisecond or less. Accordingly, the loaded power could be significantly large so that one disruption could destroy the machine. Disruption mitigation by enlarging the deposited area and lengthening the time duration is mandatory [11].

As the transient heat load, saw-tooth activity and blobs should be also considered. However, their power load would not be so large as ELM and disruptions. Recent research has suggested that the runway electrons could give significant damage [12] and should be concerned in a fusion reactor as discussed later. In addition, tokamak discharges are done basically intermittently. Consequently, power loads during plasma rump-up and rump-down phases are added to ELM heat load, and they cause thermal fatigue with power density levels of 5–20 $MW \cdot m^{-2}$ for the divertor (<1 $MW \cdot m^{-2}$ for the FW) which is a serious concern.

Material responses are divided into two categories: (i) material response to heat loads and their influence on edge plasmas (PMI), and (ii) damaging and degradation of material properties caused by plasma power load, which are described in the next section.

6.2 Material Response to Power Load and Its Influences on Boundary Plasmas

6.2.1 Spontaneous Response to Power Load

As the results of power load, various particles and photons are emitted from PFM. The direct responses of PFM to the power load are emissions of particles and photons, and they cause changes in boundary plasma, which is referred to as plasma–surface or plasma–wall interactions (PSI or PMI). The emission of charged particles including reflected and secondary ions, and electrons directly correlates to the formation of plasma sheath, which in turn modifies properties of the boundary plasma. Various photoemissions which include reflection, de-excitation of excited atoms and molecules, and radiation from heated PFM modify the boundary plasma. Higher energy photons interact with particles in the boundary plasma and consequently secondary electrons and photons are emitted. Because no plasma machines realize reactor like power load, many processes expected in reactor divertor are uncertain and remain unsolved or not understood well. High-power linear machines have been constructed or being under construction for divertor simulation [13]. However, the liner machines are hard to use as a divertor simulator, owing to the lack of strong magnetic field and rather large angle of incident particles.

As an example of phenomena not well understood, Balmer lines emission is discussed in the following. This emission is often used to determine the hydrogen recycling coefficient at PFM assuming it originated from recycled (released) hydrogen from PFS. The recycled hydrogen includes various species, reflected ions and neutrals, reemitted atoms and molecules, and some product given by chemical sputtering like water and hydrocarbons. Considering cross sections of electron excitation of these particles to be in excited states, released hydrogen flux can be estimated. However, without knowing what kinds of hydrogen particles (molecules and atoms, and those in excited states and ionized) with their energy distributions are emitted from PFS, the estimation of the recycling coefficient would include large error. In addition, as noted in Figs. 4.7 and 4.8, direct release of excited atoms from PFS also gives the Balmer line emission, which is often not considered.

6.2.2 Melting and Sublimation

When power load to PFM exceeds cooling power, the temperature of PFM easily escalates over its melting point. Generally, PFM is designed to be tolerant to steady state or normal power load. In a tokamak type fusion reactor, however, transient or off-normal power loads such as ELM and disruption seem unavoidable. Because of its huge effects, the disruption should be avoided as seldom as possible and will not be discussed here. PFM melting and sublimation under ELM appeared in current tokamaks giving significant impacts on both plasma performance and lifetime of PFM [14, 15].

Once the surface is melted, resulted liquid moves on the surface and some are released as droplets. Strong electromagnetic field in a reactor controls the melt layer motion and the direction and moving speed of the droplets. The droplets are ionized and move according to **E** x **B** force. In tokamak discharges, white small particles are often observed as line traces in the plasma. At the same time, various particles are sublimated or desorbed. If the sublimation/desorption is large enough to contaminate the plasma, plasma performance would be degraded.

When materials temperature is raised high enough, sublimation becomes appreciable. Although the sublimation removes some loaded power by the heat of sublimation, it is not large enough to cool down the PFS temperature. On the other hand, the sublimated vapor penetrates in boundary plasma followed by electrical excitation or ionization of sublimated molecules and atoms. Since the excitation/ionization energy is much larger than the heat of sublimation, it cools the edge plasma, sometimes resulting in plasma instability or disruption. However, in large tokamaks, heating power is large enough to compensate such power loss, disruption due to the sublimation is hardly observed. Instead, with aid of gyration after the ionization of sublimated atoms, most of the sublimated materials are promptly redeposited as shown in Fig. 5.3 and hardly enter plasma center. Thus, impurity transport and its accumulation in plasma center can be separated from impurity source (sublimation). Figure 6.2 is the first clear indication for that [16–18]. A Mo test limiter was exposed to TEXTOR plasma with the power load of 1.9 kW·cm^{-2} (19 MW·m^{-2}). When its temperature rose above 2100 K, the sublimation of Mo started as observed in MoI line emission. Nevertheless, other plasma parameters were hardly influenced by the Mo sublimation without the indication of central accumulation of Mo in Mo XXXVII line. This

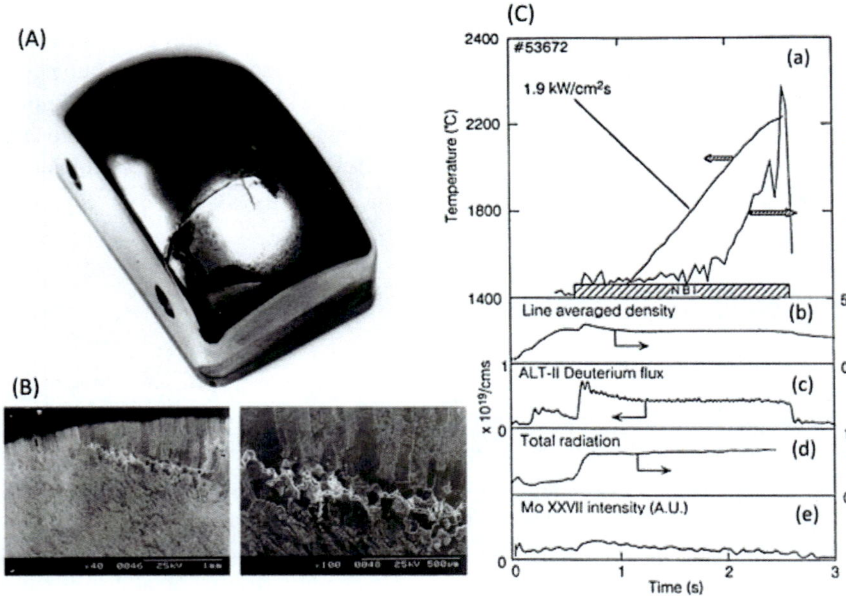

Fig. 6.2 **A** Mo test limiter after used in TEXTOR [16], **B** Cross section of Mo limiter (produced by powder metallurgy) utilized in a reversed field pinch machine (TPE-1RM15) at the Electrotechnical Laboratory, Japan. Recrystallization to columnar grains from originally granular grains with cracking between them are clearly seen and a plasma discharge accompanying the melting of Mo [17]. **C** Time evolutions of (**a**) Mo flux and surface temperature near the melting surface, (**b**) Line averaged density, (**c**) deuterium flux on the main graphite limiter, (**d**) total radiation, and (**e**) Mo XXVII line intensity in the core plasma [16]

finding motivated to use high Z materials after long period of exclusion of them as PFM in tokamaks. Now, ITER has decided to use W as armor tiles of divertor.

In large tokamaks, the power load to PFS is so large that huge amount of PFM could be sublimated. Accordingly, Carbon blooms or Beryllium blooms as appeared in high-power discharges in JET and TFTR as discussed in Sect. 6.3.1.

It should be also noted that huge amount of vapor in front of the PFS could shield succeeding plasma exposure to reduce the power load, which is referred to as vapor shielding and is expected to reduce power load to the divertor. The vapor shielding by sublimation of target materials is confirmed in linear plasma machines [19, 20], while plasma detachment near divertor target is caused by high compression or gas puffing in the divertor region. Impurity seeding like N and Ne to reduce the power load to divertor is extensively studied to use in ITER divertor [21].

6.2.3 Hydrogen Recycling

As described in Chap. 7 in detail, behavior of hydrogen injected in PFM changes with temperature. Most influential phenomena on plasma performance is reduction of hydrogen retention with temperature rise caused by power load. Because of very large incident hydrogen flux to PFM in a reactor, hydrogen concentration in near-surface region of PFM would be easily saturated. Since the saturation concentration decreases with temperature, the temperature rise enhances hydrogen release from PFM. Different from current plasma devices which are mostly operated at RT or ambient temperature, the temperature of PFM of a fusion reactor in operation should be around 700-800 K, and hydrogen moves in PFM rapidly. Therefore, even small PFM temperature rise during plasma discharge results in release of significant amount of hydrogen, which makes density control quite difficult and sometimes could lead to disruption. Another effect is wall saturation. In the current tokamak discharges, wall pumping, i.e. small hydrogen recycling or injected hydrogen being hardly released, is preferred to keep good plasma performance, or easy density control by fueling. In contrast, under higher temperature and high incident flux, hydrogen concentration in near-surface region of PFM easily comes up to saturation. This kind of wall saturation or large hydrogen recycling was attained in JT-60U of which first wall was operated around 520 K as seen in Fig. 6.3 [22, 23]. In a fusion reactor, long pulse or continuous discharge should be performed under saturated wall. Hence, it is remained as a future work to keep steady burning plasma with hydrogen saturated wall.

6.3 Damaging and Degradation of PFM

Material damages caused by very high-power loads appear in PFM as softening, recrystallization, melting, and sublimation in heating phases, and hardening, embrittlement, cracking, and fracture in cooling phases. The material lifetime of plasma-facing components (PFC) is limited mainly by material loss by erosion and the

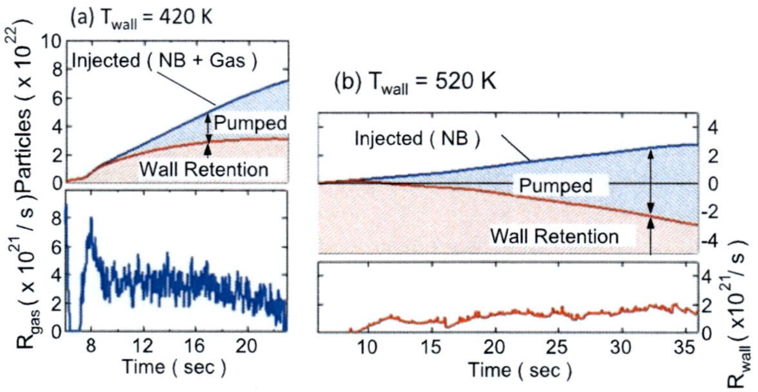

Fig. 6.3 Hydrogen wall saturation observed in JT-60U (reprinted with permission of [22]) Particles injected and retained in wall and particles flux supplied or released from wall are compared for first wall temperatures of 420 K (**a**) and 520 K (**b**). At 420 K operation, gas injection (R_{gas}) was kept to continue the discharge, while at 520 K no gas injection was required from 10 s, instead, gas was supplied by the release from the wall (R_{wall})

degradation of material properties caused by these damages. In the material damages, particularly concerned are creeps and thermal fatigues by cyclic heat load and sudden cracks caused by thermal shock load to these components. However, in normal operation, such materials failure should be avoided and are not discussed here.

Damaging and degradation are different for each PFM candidate material, in particular, C and W show totally different response to high radiation and particle loads. Therefore, both are separately described in the following sections.

To evaluate the performance of PFCs under realistic operating conditions, a complex material test program has been established that includes modeling by numerical methods, determination of all required material properties in a wide parameter range, and simulation experiments that cover the high-heat-flux performance and the neutron irradiation behavior, including synergistic effects. In this respect, high-power linear plasma machines are useful.

6.3.1 Carbon (C)

Different from metals, loaded hydrogen changes structure and chemical nature of carbon materials to hydrogen saturated carbon or hydrocarbons. However, both quite small hydrogen diffusion in carbon and chemical sputtering through hydrocarbon formation inhibit its deep diffusion and the growth of the thickness of the hydrogen saturated layers. Erosion or surface recession caused by chemical sputtering is peaked at around 800 K owing to instability of hydrocarbons at higher temperatures, while radiation enhanced sublimation (RES), mostly release of small carbon clusters, dominates the erosion above 1000 K. Therefore, material loss by erosion is large, while alteration of material properties of C by loaded hydrogen is limited within near-surface regions. Because of very high particle flux, formation of hydrogen saturated layers is immediate after the plasma exposure and afterwards response to plasma power load would become steady.

Released hydrocarbons by chemical sputtering and small C clusters by RES strongly influenced boundary plasma. Even though their significant part returns PFS by the prompt deposition, some of ionized hydrocarbons and carbon clusters enter boundary plasma and are broken up into excited/ionized hydrogen atoms, and excited/ionized carbon which in turn radiated large power. In JET and TFTR with carbon PFM, significant radiation caused by carbon was observed and referred to as carbon blooms as shown in Fig. 6.4 [24]. At around 6.7 s, Carbon density, electron density (<n_e>), Zeff and CIII emission increased, while radiation (P_{rad}) and neutron yield were decreased. In JT-60U, the strong radiation caused by carbon was observed in private flux region as carbon MARFE [25].

At that time, such high C radiation was considered undesirable. However, as described in Sect. 6.2, edge cooling by radiation is necessary and impurity seeding is planned in a fusion reactor. If C was employed as PFM, the radiation by C would enhance the boundary plasma cooling and the impurity seeding could be avoided.

In metals, injection of He is concerned, which is discussed in the next section. However, in C, because of porous nature and its inactiveness with He, injected He would be easily released resulting in little influence. Even hydrogen saturated layers are easier for He migration.

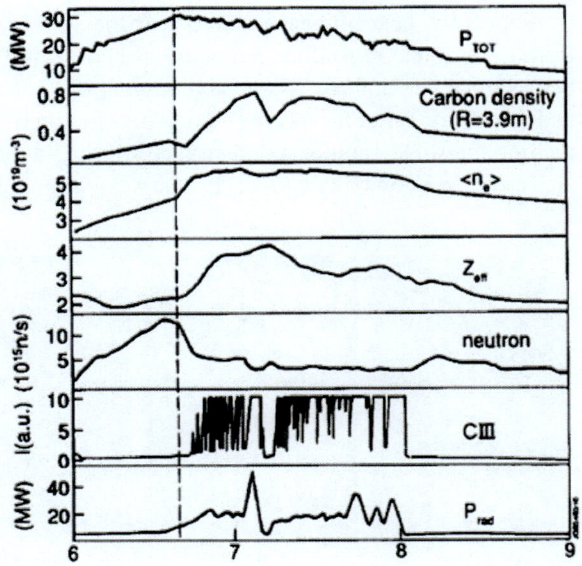

Fig. 6.4 A typical example of the effects of a carbon bloom on the plasma in JET appeared in plasma parameters; P_{TOT}: Stored energy, <n_e>: Electron temperature, Z_{eff}: Effective Z number, CIII: intensity of CIII emission, P_{rad}: Radiated power from plasma. Note the increase in carbon content and radiated power and the concurrent decrease in the fusion neutron yield (reprinted with permission from [24])

Thus, modification of C would not be significant giving little influence on boundary plasma, except carbon blooms which radiates lots and could be beneficial for edge cooling. Nevertheless, because of the concern on large T retention, C is not used in ITER. However, C could be used as plasma-facing armor tiles in a reactor [26], which is discussed in Chap. 10.

6.3.2 Tungsten (W)

6.3.2.1 Surface Damage by High-Power Load (Melting, Recrystallization, and Cracking)

Different from C, very small hydrogen solubility in W does not allow hydrogen accumulation except near-surface region, where hydrogen trapping is significant. Because of damages introduced by energetic hydrogen injection and lattice distortion given by injected hydrogen themselves, hydrogen in near-surface region accumulate to make bubbles and consequently they coalesce to make blisters. Large diffusion coefficient promotes hydrogen penetration in deep. Grain boundaries seem to attract hydrogen and exfoliation or slip of the grain boundaries are caused by the hydrogen accumulation resulting in large blisters or cleavage of small grains. This looks like hydrogen embrittlement, although the hydrogen embrittlement of W with hydrogen gas has not been ostensively shown.

Very high heat load causes temperature escalation resulting in grain growth. Consequently, heat influenced zone near the surface can be clearly distinguished from the original crystalline structure, as shown in Fig. 6.2B. Once the boundary between the heat influence zone and original one is formed, it easily triggers crack and some particles or layers are exfoliated. Only single pulsed power load even below melting threshold could result in surface cracking as shown in Fig. 6.5 [27]. Cracks are separated into two categories: primary cracks going deep along grain boundaries

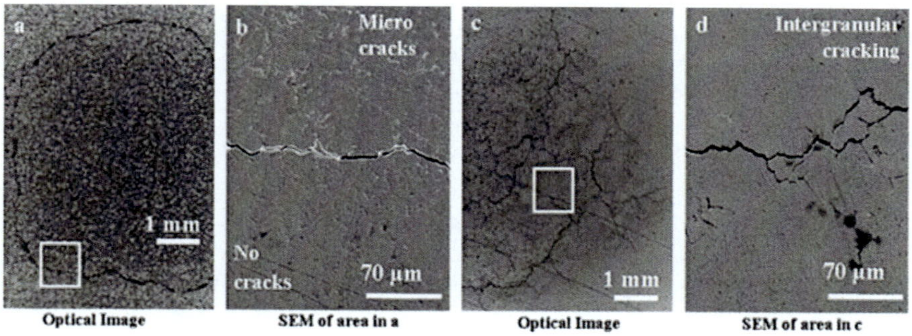

Fig. 6.5 Surface morphology and microstructures of loaded areas by single pulse of 0.55 $GW{\cdot}m^{-2}$ for 5 ms on **a**, **b** sintered W recrystallized, **c**, **d** heavily deformed W recrystallized (reprinted with permission from [27])

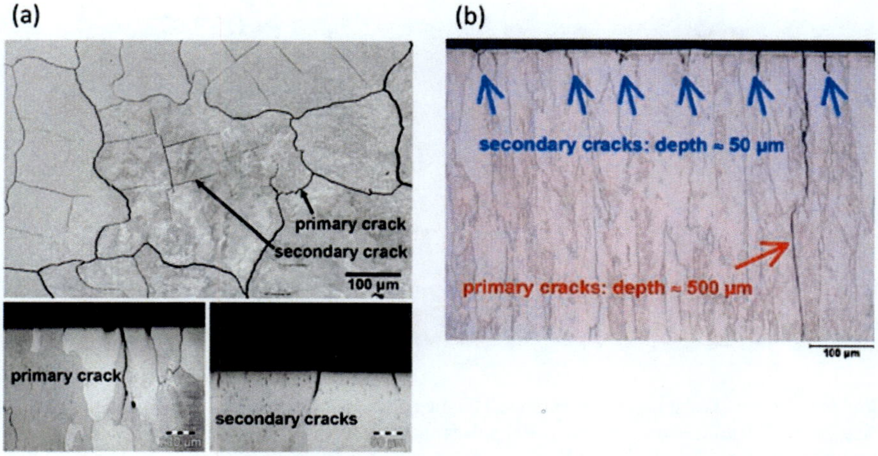

Fig. 6.6 **a** SEM image (top) of the tungsten grade M1 at the boundary between loaded and not loaded surface (P = 0.88 GW·m^{-2}, single shot)—primary and secondary cracks; cross section: LM-images of primary and secondary cracks (bottom) (reprinted with permission from [28]). **b** The cross section of the tungsten sample after 100 shots in QSPA facility with heat load of 0.9 MJ·m^{-2} and 0.5 ms time duration. Meandrous pale vertical lines are the boundaries separating elongated tungsten grains, perpendicular to the sample surface. Molten tungsten layer of 3–5 μm thickness is seen at the irradiated surface. Bold dark lines are the cracks (reprinted with permission from [29])

and secondary cracks in grains as shown in Fig. 6.6a [28]. Deep crack propagation of nearly 0.5 mm was observed in ITER like transient heat load in Fig. 6.6b [29]. The propagation of the primary crack to the boundary between PFM and the heat sink material or the cooling pipe is a serious concern in divertor structure. The cracking mostly appears in cooling phase after stopping the power load with significant influences in following discharges.

Since W is brittle near room temperature, it should be used above the ductile to brittle transition temperature (DBTT). However, because recrystallization temperature of tungsten is rather low compared to its melting point (MP), temperature window to use W as PFM is limited above ~500 K to below ~1500 K. Recrystallization increases DBTT and hence the operational temperature window would be narrowed by the repetitive power loads. Neutron damage also makes W brittle and increases DBTT. Thus, the increase of DBTT of W during the operation is one of the critical points to use W as PFM.

When the power load is high enough to melt the surface, melt layers move driven by plasma pressure. Figure 6.7 is an example of the melt layer motion caused by plasma pressure observed in heat load test in a linear plasma machine [30]. When the melt layers turn to be droplets owing to their surface tension, they depart from the surface and charge up. Then, electric–magnetic field gives the droplets motive force (**E** x **B** force) to move often observed as moving blight spots during plasma discharges. The motion varies according to the mass and the strength of the electromagnetic field.

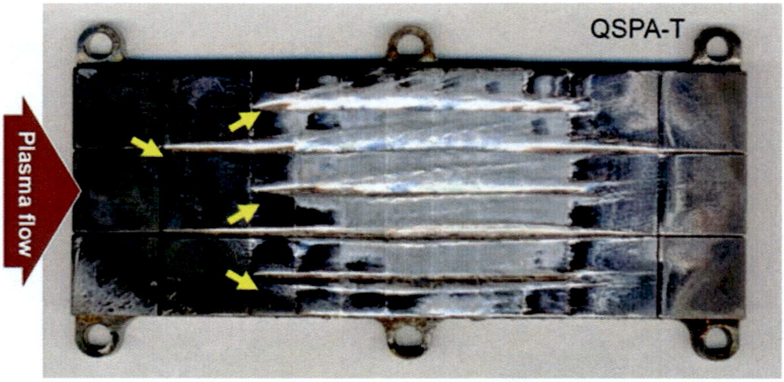

Fig. 6.7 Image of melt layer motion caused by plasma pressure observed in heat load test for ELM simulation. W surface was exposed to 100 pulses at QSPA facility ($Q_{abs} = 1.5$ MJ·m^{-2}, $\tau = 0.5$ ms) (reprinted with permission from [30])

For higher power load giving surface temperature escalation well above the melting point, massive sublimation occurs. The massive sublimation could work as vapor shielding. Nevertheless, since the incident angle of magnetic field line is very small, sublimated materials move to the mirror directly to the plasma flow and could not work as the shield. Hence, further study for the vapor shielding in divert geometry is awaited.

6.3.2.2 Surface Damage by Fuel Particles Below Melting Threshold

Blistering of W surface by hydrogen injection has been extensively studied [31–46]. Since H retention is simply given by the balance among injection, reemission and penetration, the H retention in PFS under plasma exposure can easily be over solubility limits or over saturation. In general, H concentration (C_H) equilibrated with H_2 gas pressure ($P(H)_2$) is given by $C_H = S{\cdot}P(H_2)^{1/2}$, where S is H solubility. Hydrogen injected over the saturation is either in trapped or super solution states, and it is self-stabilized to make bubbles or self-trapping sites. The bubbles accumulate to blisters when their internal gas pressure exceeds mechanical strength of PFM, GPa order [46]. In this gas pressure mechanism, in principle, blisters appeared in grains are dome shaped and there is no difference in the mechanism of blister formation by H and He implantations and the thickness of blister skins is nearly the same as the implanted depth (projected range) of H and He, respectively.

In the case of W, thickness of the blister skins is often much larger than the projected range of injected H, suggesting the thickening of the H saturated layers over the projected range. Sometimes large blisters over grain sizes were observed as shown in Fig. 6.8 [37], while the radius of dome-shaped blisters is usually smaller than the grain size in Fig. 6.9 [38]. Such large blisters or sometimes exfoliation over grains are caused by different mechanisms than the gas pressure mechanism.

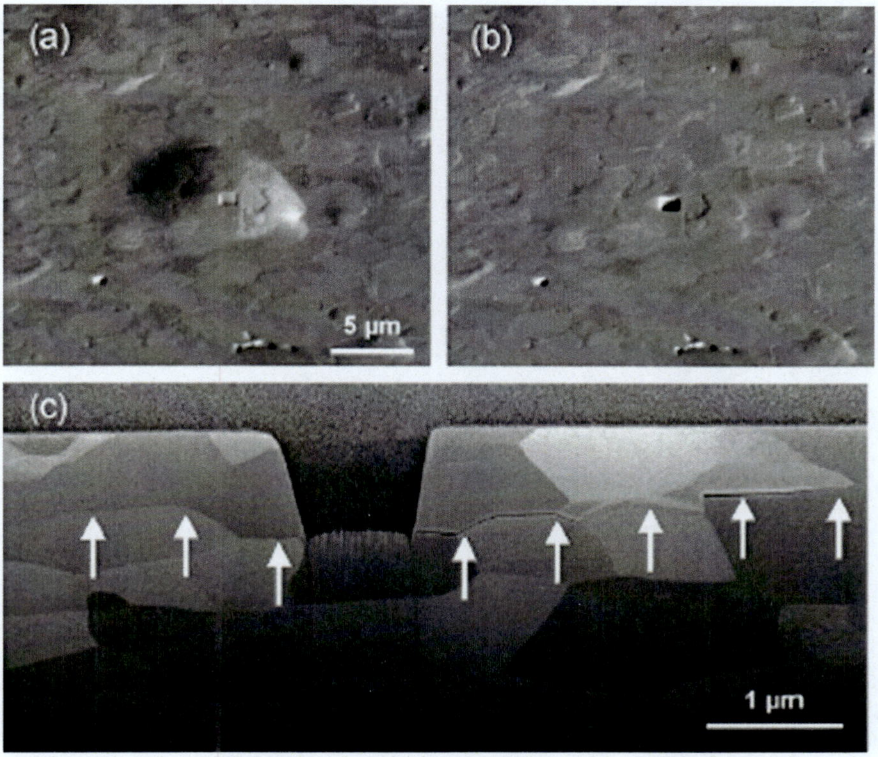

Fig. 6.8 **a** SEM image of an individual blister on polycrystalline tungsten after plasma exposure (38 eV/D, 3 × 10^{24} D·m-^{2}, 300 K) and **b** same surface area after sputtering a hole with Focused Ion Beam (FIB) into the blister cap and its elastic collapse after degassing and **c** cross section through the cap of the collapsed blister at the position of the hole sputtered to degas the blister. Arrows indicate the remaining crack system of the former blister (**a**) and (**b**) tilted by 52°, (**c**) by −38° (reprinted with permission from [37])

As already described, hydrogen is trapped and accumulated at grain boundaries. It weakens the bonding between neighboring grains. In addition, H inclusion in grains gives swelling and, as a consequence, introduces lateral stress. Since W materials are usually manufactured by the powder metallurgy (PM) including hot rolling, their grains tend to extend parallel to the surface. Accordingly, a particular grain parallel to the surface is weakened by H segregated at its edge and exfoliated from the matrix to make dome-shaped blisters [38, 41] (gas pressure mechanism) or convexity (or ledges) [33, 35, 38] due to slipping along specific crystalline planes together with exfoliation of grain boundaries (stress mechanism) as shown in Fig. 6.9B [38].

To clarify the blister mechanism [43, 46], it is important to know whether blister skins are permanently deformed (plastic deformation) or elastically deformed. Even such larger blisters than grain sizes were produced by the elastic deformation and

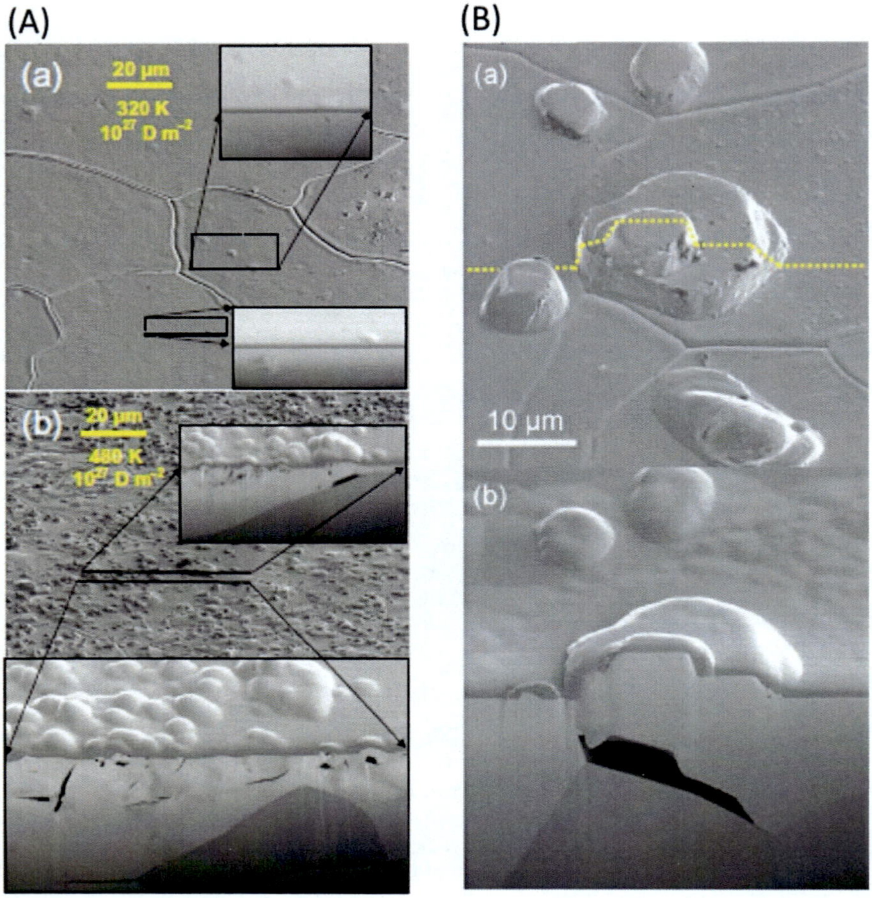

Fig. 6.9 **A** Tungsten surface exposed to 10^{27} Dm^{-2} at 320 K, surface tilted by 52°(**a**) and at 480 K, tilted by 70°(**b**). The inserts in (**a**) show the cross section through one of the sparse blisters as well as through a distorted region below the surface, which does not exhibit a noticeable blister at the surface. Inserts in (**b**) show strongly distorted and cracked regions, the upper overlay shows also a cavity at grain boundary beneath the large blister-like structure. The bold lines indicate the positions of the cross sections with marked surface area shown in inserts (all inserts tilted by 52°. Surface in inserts was coated by Pt–C film. **B** high-dome blister (**a**) before and (**b**) after cross sectioning with FIB (surface tilted by 52°; cross section by −38°. Line in (**a**) indicates the position of cross sectioning (**b**). The surface in (**b**) is coated with a mixed Pt–C film (reprinted with permission from [38])

returned to its original flat shape without any traces of the blister by the release of gas inside [37] (see Fig. 6.8). It is also clear that the convexity owing to the slipping is caused by plastic deformation [38–41].

Applying grazing incidence electron microscopy (GIEM) [47], Enomoto et al. [46] have clearly shown that dome-shaped blisters appeared on single crystalline W by H, D, and He ion implantations were permanently deformed after higher fluence

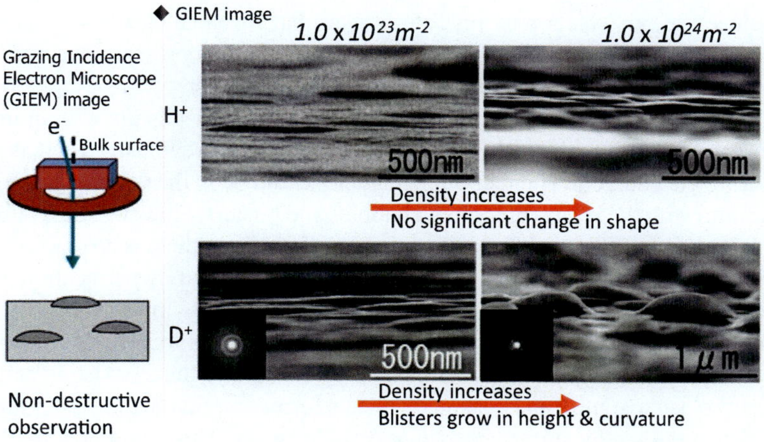

Fig. 6.10 Observation of blisters produced by 1.5 keV H^+ and D^+ ion irradiation on W single crystalline at 400 K by glazing angle incidence electron microscopy (GIEM) [47]. The diffraction patterns given in both images as in sets show a hallow patter indicating blister skins being amorphous structure for low fluence and diffraction spots indicating recrystallized to very fine grains

of irradiation as shown in Fig. 6.10 [38]. A diffraction pattern, see as the inset of Fig. 6.10, obtained from the blister skins at lower fluence indicates that blister skins were amorphized, while at higher fluence the skins consisted of very fine crystalline grains (nano-grains). This indicates that the elastically formed blister changes to permanently deformed blisters.

There is no doubt that the deformation is caused by mechanical stress irrespective of the cause, either by gas pressure or lateral stress. Therefore, thickness and radius of dome-shaped blister can be related to mechanical properties of hydrogen saturated layers of tungsten, but not pure W. And yield stress of the layers or blister skins (Y) can be correlated with thickness (t) and radius (r) of the blisters and gas pressures (P) as

$$P = 4Yt/r. \tag{6.1}$$

From separate EELS (Electron Energy Loss Spectroscopy in GIEM) measurements of gas pressure in the bubbles, the yield stress of the H saturated layers was found to be raised by two or three times from that of the pure W [47]. This is very much consistent with the increase of hardness and DBTT of W by ion irradiation. The Eq. 6.1 also indicates that smaller the radius of the blister, the gas pressure or lateral stress to make the bubble is larger. In this respect, it is interesting to note that Balden et al. [37] have shown elastically formed blisters. They have found that the dome-shaped large blister retained H_2 gas inside and drilling a hole in the blister skins resulted in gas release and the disappearance of the blister (see Fig. 6.8). In this particular case, the blister radius is larger than the grain size, and the stress to produce the blister must be much less than that for small blisters observed on the single crystal. In another word, the lateral stress or gas pressure overcomes the bonding force between grains to form such large size blister.

A question arises; why hydrogen can stay in bubbles or blisters with quite high pressure in spite of large diffusion coefficient as discussed previously. It is reasonable that during the energetic H injection, H fugacity could be in the GPa range and be in equilibrium with the H_2 gas pressure in bubbles and blisters. However, the H fugacity must decrease at the termination of the injection. Therefore, there is no reason to keep H_2 gas pressure in the blisters high. H_2 gas could be released by permeation, except for permanently deformed blisters. One possible answer is impurity deposition or accumulation on the inner surface of the blister skin to prohibit H release from the blister inside. The observation by Shimada et al. [48] states that coexisting carbon in hydrogen plasma significantly enhance blister formation that suggests suppression of hydrogen release by surface carbon or carbide layers. This could potentially also happen on bubble surface preventing hydrogen in bubbles from escape by some impurities covering their inside surfaces. Another possibility is that once hydrogen concentration in W is raised above a certain level making the H saturated layers, hydrogen diffusion in it is likely suppressed. Considering all observations described above, hydrogen accumulation in W during the H irradiation was modeled as shown in Fig. 6.11 and discussed in more detail in Chap. 9 [46].

He injection makes surface blisters. He hardly diffuses in W, and is easily accumulated into small bubbles. Aggregation of small bubbles in near-surface regions

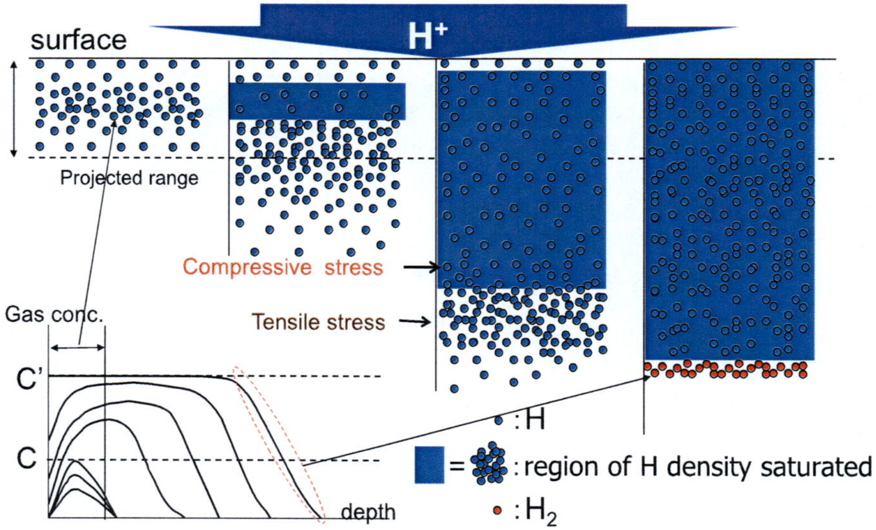

Fig. 6.11 Expected H depth profile to bring blistering in W [46]. With increasing the injected fluence of H, the thickness of H saturated layers grows over projected range. Some volume expansion of the H saturated layers, compress stress, and tensile stress are caused at the boundary and could result in blistering and/or exfoliation of the H saturated layers (see also Fig. 9.9)

results in dome-shaped He blister. In case of He injection at higher temperature, not only He but also vacancies produced by He injection and they move to the surface. Accordingly, characteristic structure referred to as the fuzz structure is formed as noted in Sect. 4.6.2. However, such structure would not be stable under high-power load on PFS, because heat transfer through its needle-like structure could not be high enough to avoid the melting. Once the fuzz structure is formed whatever the causes are, its melting and evaporation influence the edge plasma. Researches on this point are on-going.

6.3.3 Other PFM Candidates (Be and Li)

It used to be quite important to reduce oxygen impurity in plasma and materials working as oxygen getters, like Li, Be, Al, and Ti have been PFM candidates. Be is used in ITER like wall in JET considering its use in ITER. However, its low melting point would not allow to use in a reactor. Be as a metal would behave like W, brittle and having quite low hydrogen solubility and diffusivity. Different from W, Be reacts with hydrogen to make volatile hydride, BeH_2, and chemical sputtering of Be by H like C could result in large erosion [49, 50]. However, BeH_2 is not stable at higher temperature, and erosion of Be in JET ITER-like wall is not likely significant. Because of their low melting points, other PFM candidates except Li will not be used in a reactor. A concept to use liquid Li wall attracts much interest, because Li works as both coolant and breeder [51–54]. Except significant evaporation by high-power load, material modifications occurred in solid materials like radiation damages are not concerned in liquid. Evaporated Li would return to PFM by gyration after ionization. Furthermore, massive evaporation would work as vapor shielding to reduce the power load. Hydrogen gettering effect with forming LiH which stays as tiny solid particles in liquid Li reduces hydrogen recycling and realizes wall pumping to make plasma density control easier. However, large wall pumping makes in both T fuel throughput and T retention in Li very large. Hence quick recovery of T from Li is critically important to use liquid Li as PFS, which is unfortunately not easy because LiH particles dispersed in Li is quite hard to recover.

6.3.4 Structure Materials

As a structure material, iron-based steels are most reliable. Particularly, low activation ferrite martensitic steels are the candidate first wall or structure material. Although martensitic steels are another candidate, their toughness at higher temperature is concerned to use them in a reactor. Unfortunately, melting points of these iron-based steels are not high enough to use them as PFM and some protective layers are necessary. In most of present reactor designs, W is employed as armor tiles and the tiles should be joined to a heat sink or structure material like the steels or Cu alloys. Since the volume expansion coefficient of W is much less than these materials, adhesion in the joining is not easy. Furthermore, cyclic heat load or heat shock could destroy the joint. To make ITER divertor with W armors on a heat sink material is one of the most difficult engineering tasks. Carbon materials, if they will be used as PFM, have the similar problems in brazing to the heat sink material. In addition, the carbon materials will swell or expand volumetrically due to neutron irradiation, which also gives large stress on the joint.

Because of concerns on large T retention, C was excluded to use in ITER. Nevertheless, the use of carbon tiles as armor is beneficial to reduce T permeation in structure materials comparing to W armor, because most of the injected hydrogen will retain in near surface and not migrate deep [26].

6.4 Summary

The power load to PFS appears as plasma materials interactions (PMI), which are similar to what the surface of a rocket running into the sun would be exposed to and, at present, the physical and chemical phenomena expected in PMI in a fusion reactor would be very difficult to study directly.

In this chapter, after introduction of estimated power load to PFM in a fusion reactor,

material response to the power load which influences plasma is summarized. Spontaneous response of PFM to the power load appears as emission of particles and photons, including sputtered atoms and ions of PFM and surface impurities, sublimated atoms, secondary electrons and photons, and radiation from PFM. Among all, sputtered materials would give significant influence on plasma.

Owing to power load given by the energetic photons and particles, surface and subsurface of PFM are modified or damaged, sometimes resulting in surface melting or massive sublimation. Although these surface modifications are not spontaneous, they gradually change materials properties which in turn changes response to the power load of PFM. Since the power load is given by energetic photons (radiation) and energetic fuel (H) particles, fuel recycling between plasma and PFM also significantly influence plasma confinement. The material modification influences behavior of fuel in PFM, resulting in changes in fuel recycling. Hence, material response to the power load is synergistic of heating, modification or damaging of PFM, and the fuel behavior, which makes understanding of PMI quite difficult.

Among presently operating plasma apparatus, only JET can realize similar or a little lower levels of power load, therefore, understanding of PMI in a fusion reactor must be extrapolated from observations on the currently operating apparatus and laboratory experiments for basic understanding of individual processes of damaging and hydrogen recycling.

In the last half of this chapter, damaging and degradation of Carbon and Tungsten (W) are separately introduced with consideration of their influence on hydrogen recycling. Li and Be are briefly summarized as PFM. Although structure materials like stainless steels should not be used as PFM, hydrogen recycling could be influenced by those materials used as heat sink of PFM.

It should be mentioned again that PMI is a critical scientific issue for establishment of a fusion reactor as a power source, with major potential limitations on plasma core and edge operating parameters. Gaining understanding and predictive capability in this area will require simultaneously addressing complex and diverse physics and chemistry occurring over a wide range of lengths (angstroms to meters) and times (femtoseconds to days).

References

1. J. Linke, T. Hirai, M. Rödig et al., Performance of plasma-facing materials under intense thermal load in tokamaks and stellarators. Fusion Sci. Technol. **46**, 142–151 (2004)
2. N. Asakura, K. Hoshino, S. Suzuki et al., Studies of power exhaust and divertor design for a 1.5 GW-level fusion power DEMO. Nucl. Fusion **57**, 126050 (2017)
3. R. Wenninger, R. Albanese, R. Ambrosino et al., The DEMO wall load challenge. Nucl. Fusion **57**, 046002 (2017)
4. V. Barabash, G. Federici, J. Linke et al., Material/plasma surface interaction issues following neutron damage. J. Nucl. Mater. **313–316**, 42–51 (2003)
5. Yu.F. Baranov, C.D. Challis, J. Ongena et al., Large ELM-like events triggered by core MHD in JET advanced tokamak plasmas: impact on plasmas profiles, plasma-facing components and heating systems. Nucl. Fusion **52**(14), 023018 (2012)
6. R.A. Pitts, G. Arnoux, M. Beurskens et al., The impact of large ELMs on JET. J. Nucl. Mater. **390–391**, 755–759 (2009)
7. A. Huber, G. Arnoux, M.N.A. Beurskens, Radiation loads onto plasma-facing components of JET during transient events–experimental results and implications for ITER. J. Nucl. Mater. **415**, S821–S827 (2011)
8. V. Rohde, M. Balden, the ASDEX Upgrade Team, Arc erosion of full metal plasma facing components at the inner baffle region of ASDEX Upgrade. Nucl. Mater. Energy **9**, 36–39 (2016)
9. H. Timko, F.F. Djurabekova, K. Nordlund et al., Mechanism of surface modification in the plasma-surface interaction in electrical arcs. Phys. Rev. B **81**, 184109 (2010)
10. D.L. Rudakova, C.P. Chrobak, R.P. Doerner et al., Arcing and its role in PFC erosion and dust production in DIII-D. J. Nucl. Mater. **438**, S805–S808 (2013)
11. T.E. Evans, ELM mitigation techniques. J. Nucl. Mater. **438**, S11–S18 (2013)
12. A.H. Boozer, Runaway electrons and ITER. Nucl. Fusion **57**(14), 056018 (2017)
13. A. Kreter, Reactor-relevant plasma-material interaction studies in linear plasma devices. Fusion Sci. Technol. **59**, 51–56 (2011)
14. J.W. Coenen, G. Arnoux, B. Bazylev et al., ELM induced tungsten melting and its impact on tokamak operation. J. Nucl. Mater. **463**, 78–84 (2015)
15. G. Sergienko, B. Bazylev, T. Hirai, Experience with bulk tungsten test-limiters under high heat loads: melting and melt layer propagation. Phys. Scr. **T128**, 81–86 (2007)
16. T. Tanabe, V. Philipps, B. Unterberg et al., High Z limiter test in TEXTOR. J. Nucl. Mater. **212–215**, 1370–1378 (1994)
17. T. Tanabe, M. Fujine, H. Noguchi et al., High heat load test of molybdenum. J. Nucl. Mater. **200**, 120–127 (1993)

18. T. Tanabe, Y. Ueda, V. Philipps et al., High Z limiter test in TEXTOR: thermal response and post-mortem analysis. Fusion Eng. Des. **28**, 13 (1995)
19. D.I. Skovorodin, A.A. Pshenov, A.S. Arakcheev, Vapor shielding models and the energy absorbed by divertor targets during transient events. Phys, Plasmas **23**, 022501 (2016)
20. V.M. Safronov, N.I. Arkhipov, I.S. Landman, Evaporation and vapor shielding of CFC targets exposed to plasma heat fluxes relevant to ITER ELMs. J. Nucl. Mater. **386–388**, 744–746 (2009)
21. A. Kallenbach, M. Bernert, R. Dux et al., Impurity seeding for tokamak power exhaust: from present devices via ITER to DEMO, Plasma Phys. Control. Fusion **55**(10), 124041 (2013)
22. T. Nakano, N. Asakura, H. Takenaga et al., Impact of wall saturation on particle control in long and high-power-heated discharges in JT-60U. Nucl. Fusion **46**, 626–634 (2006)
23. T. Nakano, N. Asakura, H. Takenaga et al., Particle balance under global wall saturation in long-pulse discharges of JT-60U. J. Nucl. Mater. **363–365**, 1315–1322 (2007)
24. M. Ulrickson, The JET team, A review of carbon blooms on JET and TFTR. J. Nucl. Mater. **176–177**, 44–50 (1990)
25. T. Nakano et al., Radiation process of carbon ions in JT-60U detached divertor plasmas. J. Nucl. Mater. **390–391**, 255–258 (2009)
26. T. Tanabe, Revisiting Carbon Materials as Plasma-Facing Materials of a Fusion Reactor. Plasma Phys. Rep. **45**, 300–314 (2019)
27. I. Uytdenhouwen, M. Decréton, T. Hirai et al., Influence of recrystallization on thermal shock resistance of various tungsten grades. J. Nucl. Mater. **363–365**, 1099–1103 (2007)
28. G. Pintsuk, W. Kühnlein, J. Linke, M. Rödig, Investigation of tungsten and beryllium behavior under short transient events. Fusion Eng. Des. **82**, 1720–1729 (2007)
29. S.E. Pestchany, J. Linke, Simulation of cracks in tungsten under ITER specific transient heat loads. Fusion Eng. Des. **82**, 1657–1663 (2007)
30. A. Zhitlukhin, G. Federici, R. Giniyatulin, et al. Experimental assessment of the effects of ELMs and disruptions on ITER divertor armour materials, in *Proceedings of 20th IAEA Fusion Energy Conference, 2004*, IT/P3–30
31. V Kh, Alimov, J. Roth, R.A. Causey R A et al., Deuterium retention in tungsten exposed to low-energy, high-flux clean and carbon-seeded deuterium plasmas. J. Nucl. Mater. **375**, 192–201 (2008)
32. Y. Ueda, M. Fukumoto, J. Yoshida et al., Simultaneous irradiation effects of hydrogen and helium ions on tungsten. J. Nucl. Mater. **386–388**, 725–728 (2009)
33. V.Kh. Alimov, W.M. Shu, J. Roth et al., Temperature dependence of surface topography and deuterium retention in tungsten exposed to low-energy, high-flux D plasma. J. Nucl. Mater. **417**, 572–575 (2011)
34. G.M. Wright, M. Mayer, K. Ertl et al., Hydrogenic retention in irradiated tungsten exposed to high-flux plasma. Nucl. Fusion **50**, 075006 (2010)
35. V.Kh. Alimov, W.M. Shu, J. Roth et al., Surface morphology and deuterium retention in tungsten exposed to low-energy, high flux pure and helium-seeded deuterium plasmas. Phys. Scr. **T138**(5), 014048 (2009)
36. A. Manhard, K. Schmid, M. Balden, W. Jacob, Influence of the microstructure on the deuterium retention in tungsten. J. Nucl. Mater. **415**, S632–S635 (2011)
37. M. Balden, S. Lindig, A. Manhard, J-Ha You, D_2 gas-filled blisters on deuterium-bombarded tungsten. J. Nucl. Mater. **414**, 69–72 (2011)
38. S. Lindig, M. Balden, V.Kh. Alimov, et al. Subsurface morphology changes due to deuterium bombardment of tungsten, Physica Scripta, **T138** (2009) 014040(5 pp)
39. W.M. Shu, G.-N. Luo, T. Yamanishi, Mechanisms of retention and blistering in near-surface region of tungsten exposed to high flux deuterium plasmas of tens of eV. J. Nucl. Mater. **367–370**, 1463–1467 (2007)
40. W.M. Shu, A. Kawasuso, T. Yamanishi, Recent findings on blistering and deuterium retention in tungsten exposed to high-fluence deuterium plasma. J. Nucl. Mater. **386–388**, 356–359 (2009)
41. W.M. Shu, M. Nakamichi, VKh Alimov et al., Deuterium retention, blistering and local melting at tungsten exposed to high-fluence deuterium plasma. J. Nucl. Mater. **390–391**, 1017–1021 (2009)

42. T. Hoshihira, T. Tanabe, T. Otsuka, Visualization of hydrogen distribution around blisters by tritium radio-luminography. J. Nucl. Mater. **386–388** 776–779 (2009)
43. T. Hoshihira, T. Otsuka, R. Wakabayashi, T. Tanabe, Hydrogen behavior near surface regions in Mo and W studied by tritium tracer technique. J. Nucl. Mater. **417**, 559–563 (2011)
44. T. Shimada, T. Funabiki, R. Kawakami et al., Carbon behavior on tungsten surface after carbon and hydrogen mixed beam irradiation. J. Nucl. Mater. **329–333**, 747–751 (2004)
45. T. Tokunaga, M.J. Baldwin, R.P. Doerner et al., Blister formation and deuterium retention on tungsten exposed to low energy and high flux deuterium plasma. J. Nucl. Mater. **337–339**, 887–891 (2005)
46. N. Enomoto, S. Muto, T. Tanabe, Grazing-incidence electron microscopy of surface blisters in single- and polycrystalline tungsten formed by H+, D+ and He+ irradiation. J. Nucl. Mater. **385**, 606–614 (2009)
47. S. Muto, T. Matsui, T. Tanabe, Observation of surface blistering by grazing incidence electron microscopy, Japan. J. Appl. Phys. **39**, 3555–3556 (2000)
48. T. Shimada, H. Kikuchi, Y. Ueda et al., Blister formation in tungsten by hydrogen and carbon mixed ion beam irradiation. J. Nucl. Mater. **313–316**, 204–208 (2003)
49. R.P. Doerner, M.J. Baldwin, D. Buchenauer, The role of beryllium deuteride in plasma-beryllium interactions. J. Nucl. Mater. **390–391**, 681–684 (2009)
50. S. Brezinsek, JET-EFDA contributors, Plasma-surface interaction in the Be/W environment: conclusions drawn from the JET-ILW for ITER. J. Nucl. Mater. **463**, 11–21 (2015)
51. J.P. Allain, C.N. Taylor, Lithium-based surfaces controlling fusion plasma behavior at the plasma-material interface. Phys. Plasmas **19**, 056126 (2012)
52. Y. Nagayama, Liquid lithium divertor system for fusion reactor. Fusion Eng. Des. **84**, 1380–1383 (2009)
53. R.E. Nygren, F.L. Tabarés, Liquid surfaces for fusion plasma facing components—A critical review. Part I Phys. PSI Nucl. Mater. Energy **9**, 6–21 (2016)
54. R.J. Goldston, A. Hakim, G.W.Hammett et al., Recent advances towards a lithium vapor divertor. Nucl. Mater. Energy **12**, 1118–1121 (2017)

Chapter 7
Fundamentals of Hydrogen Recycling

7.1 Introduction

Hydrogen recycling at plasma-facing surface (PFS) is quite important for density control or maintaining plasma in good condition. In present tokamaks, low recycling scheme, i.e. most of impinging hydrogen to plasma-facing materials (PFM) being retained with little reemission (referred to as wall pumping), is favored for easy density control by fueling only. In order to attain the large wall pumping, in most of the present tokamaks, "wall conditioning" to remove hydrogen retained in PFM such as He glow discharges has been routinely made after certain numbers of main plasma discharges [1]. However, a long pulse or steady-state operation in a fusion reactor would not keep the low recycling scheme, and fuel accumulation in PFM continues to increase up to the saturation of fuels in PFM. In addition, impinging hydrogen flux to PFS in a reactor is so high that retention would be saturated within short discharge time. This means that in a fusion reactor, plasma operation in high recycling scheme is dispensable, which is not well established yet and one of the important research targets in ITER.

Even after the surface saturation, hydrogen diffuses into deep to increase T retention continuously during the plasma operation, which are concerns for T safety and fuel self-sufficiency. Although techniques developed for the wall conditioning could be used to reduce the T inventory, processing or refinement of removed T is required. For example, He grow discharge results in huge amount He gas containing T, which gives undesired load on the refinement system of exhausted fuels.

Therefore, investigation and understanding of H recycling and its influence on plasma confinement are urgent tasks. However, no tokamaks having high-temperature PFS and high H flux are currently available. Although linear plasma apparatus having high H flux are available, they are mostly used for high heat flux tests of PFM and do not seem suitable for simulation of the H recycling in a reactor divertor owing to their different geometry from that of a rector divertor and lack of magnetic field. Therefore, currently available data on H recycling should be extrapolated to reactor environment. This chapter is devoted to summarize current knowledge and

T. Tanabe, *Plasma-Material Interactions in a Controlled Fusion Reactor*, Springer Series in Plasma Science and Technology, https://doi.org/10.1007/978-981-16-0328-0_7

understanding of hydrogen recycling in tokamaks and those given by laboratory experiments. Fundamentals of hydrogen interactions with fusion reactor materials are written in a textbook [2].

7.2 Overall Fuel Flow at Steady-State Burning

Figure 7.1 shows a schematic diagram of D and T fuel flow at steady-state burning. Fuel throughputs of D and T with respective fluxes of ϕ_D, ϕ_T (g/s) will be controlled separately to optimize burning efficiency. In the tokamak plasma, fueling efficiency, i.e. the ratio of fuels going into the plasma and the throughput, is not large, usually around 10–30% or less. Furthermore, burning of the fuel going into the plasma is also small, most probably 5% or less [3]. In the figure, the fueling efficiencies are given by f_D and f_T, and burning efficiency by b. The DT reaction produces 14 MeV neutron and 3.5 MeV He. The former is converted to electric or thermal power and at the same time used for production of T in the blanket, and the latter for heating to sustain burning plasma. Thermalized He must be evacuated to avoid the dilution of the fuels in the burning plasma.

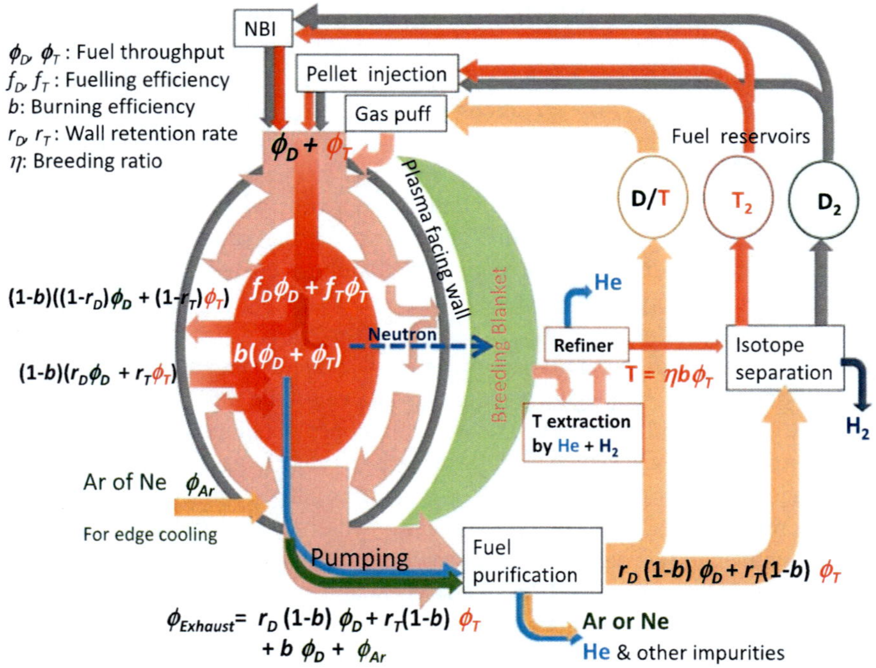

Fig. 7.1 Schematic diagram of D and T fuel flow at steady-state burning. ϕ_D, ϕ_T: Fuel throughput, ϕ_{Ar}: Impurity seeding throughput, f_D, f_T: Fueling efficiency, b: Burning efficiency, r_D, r_T: Wall retention rate, η: Breeding ratio

For enhancement of radiative cooling of the burning plasma and also for disruption mitigation, impurity seeding such as Ne and Ar will be employed [4], which should be also exhausted. In addition, various chemical forms of hydrocarbons will be formed from carbon if used as a plasma-facing material, and H_2 and H_2O always remains as residual gas in a vacuum system. Thus, the most of D and T fueled are exhausted as a heavily contaminated gas with H, H_2O, various hydrocarbons and inert gasses (He, Ne, and Ar).

Some fuels are escaping from plasma and impinge into PFM and some are retained. The recycling coefficient (r) defined as the flux ratio of incident and outgoing fuels at PFS gives the fuel retention rate as (1-r). Neutrons are going to blanket systems to breed T with a breeding ratio, η, defined as the ratio of the number of produced neutrons to that of bred T atoms. The breeding ratio, which is strongly influenced by geometrical structure, breeding materials, neutron multipliers of the blanket system, does not likely exceed 1.2.

Hydrogen recycling coefficient, r, is not necessarily constant. It changes when incident fluxes of the fuels and impurities (He ash, residual gases like water, and seeded gases like Ne and Ar) and material's temperature. Changes of energy of incident fuel particles also force to change r. In the following sections, a little detail of fuel (referred to as hydrogen) recycling or behavior of hydrogen particles escaping from plasma with higher energy is described following general features of hydrogen recycling at PFM (metals) shown in Fig. 7.2 [5]. Energetic hydrogen ions and neutrals escaping from plasma impinge to PFS. Except directly reflected ones which are a few tens % of the impinging flux, the hydrogen ions and neutrals are injected into the material within a certain depth referred to as projected range (varied depending on incident energy). Most of the injected hydrogen particles diffuse back to the front (injected) surface to be reemitted (reemission), and remaining migrates or diffuses

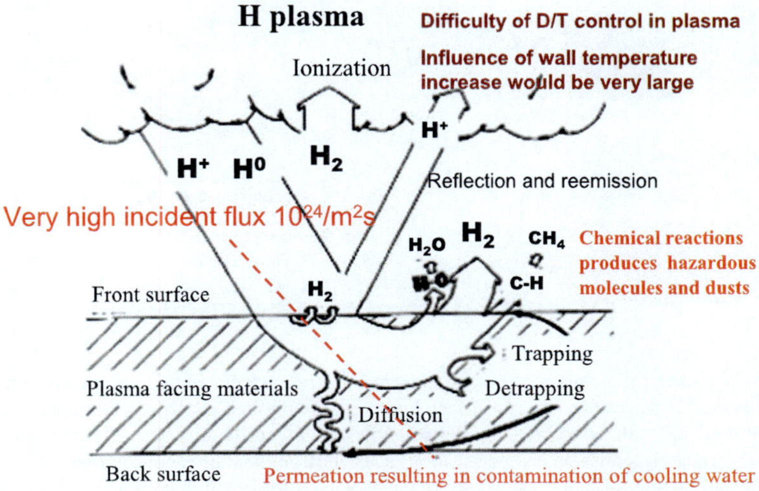

Fig. 7.2 Schematic diagram of fuel (Hydrogen) flow at/in PFM

into be dissolved or trapped and some permeate to the back surface to be released as molecules of T_2, DT, D_2, HT, and DT. In the following sections, injection of energetic hydrogen, reflection and remission, permeation and hydrogen retention (fuel inventory) are separately introduced.

7.3 Injection of Energetic Hydrogen

Figure 7.3 schematically shows potential energy diagram for energetic hydrogen injection in a metal. In a fusion reactor, high-energy ions or neutral atoms which are produced by charge exchange process of neutral particles with ions are injected to PFS. In addition, molecules and atoms in excited states, and ions in boundary plasma accelerated with sheath potential are injected. Since dissociation energy of H_2 is 4.7 eV and ionization energy of H, 13.6 eV, H atoms and ions are in higher energy states and penetrate directly into subsurface layers over surface barrier even if it exists. As an extreme case, when incident hydrogen have very high-energy, some of them can permeate through to the back surface, which is sometimes referred to as super-permeation. Once energetic hydrogen are injected in, they continue to move

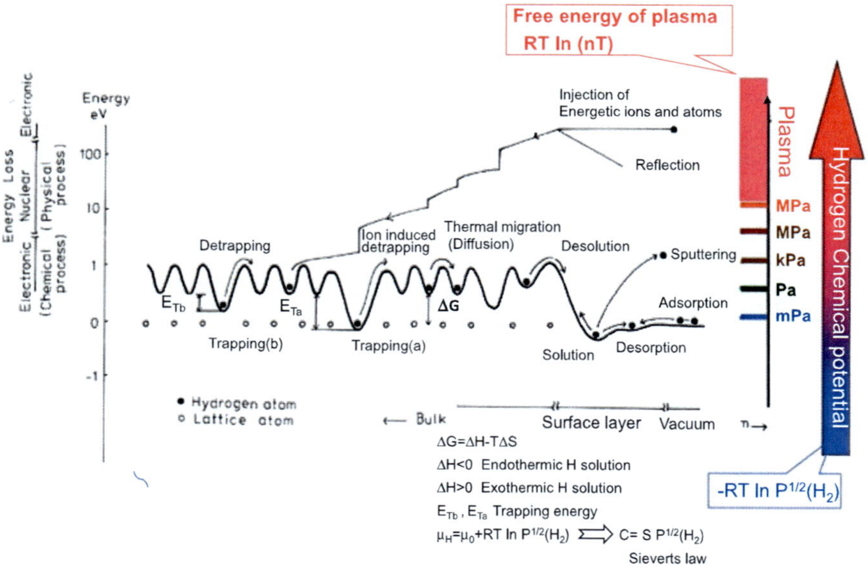

Fig. 7.3 Schematic illustration of potential energy diagram for energetic hydrogen injection in a metal. Injected H penetrates into subsurface losing energy through electron excitation and nuclear collision until thermalized to be target temperature. Afterward, H migrates/diffuses to both front and back surfaces (reemission and permeation) caused by gradient of H concentration or chemical potential. During the migration, some are trapped. Some of injected H collides with dissolved or trapped H resulting in ion-induced diffusion, detrapping, and desorption

until they are thermalized to the wall temperature. Afterwards they diffuse into deep or back to the injected surface according to the gradient of chemical potential or concentration of H in the wall. During the diffusion, some of them are trapped.

Energy loss processes of high energetic hydrogen injected in a metal are twofold, one is energy loss by excitation of electrons in the metal (electron energy loss, or electron stopping), the other is nuclear collision (nuclear stopping). If the energy of the hydrogen particles is high enough to displace a lattice atom, the lattice atom is displaced remaining a Frenkel pair, i.e. an interstitial atom and a vacancy. If the incident energy is not large, the nuclear collision occurs at surface resulting in sputtering and surface recession. Most of the Frenkel pars are recombined. However, some remain as defects and defect clusters, such as stacking fault, interstitial lops, vacancy clusters, vacancy loops, and so on (see Sect. 4.2). Usually, these defects trap hydrogen to be immobilized. In Chap. 8, hydrogen trapping in various defects of plasma-facing candidate materials are described in detail.

Owing to hydrogen trapping in defects produced by injected hydrogen themselves, implanted hydrogen diffusion or migration is strongly influenced by the newly introduced defects. Since the depth of the defects induced is a little shallower than the implanted hydrogen depth, the implanted hydrogen move to the defects resulting in the depth profile of the implanted hydrogen being shallower than the implanted profile. Since the amount of trapped hydrogen is far larger than that of dissolved hydrogen, hydrogen detrapping dominates in hydrogen recycling at lower temperature and wall pumping could continue long. However, trapping decreased with increasing temperature which makes low recycling difficult for high-temperature wall.

In a reactor vessel, residual hydrogen gases escaping from plasma are always existing with pressure P (around 1–100 Pa), and dissolve into PFM following the Sievert's law with concentration c given by

$$c = S \cdot P^{1/2}, \tag{7.1}$$

where S is the Sievert's constant with temperature dependence of

$$S = S_0 \exp(-Es,/RT), \tag{7.2}$$

where S_0, Es, R, and T are a constant, heat of solution, gas constant, and temperature (K).

Although S increases with temperature in most of the metals except hydride forming ones, H uptake in metals from the residual fuel gas is much less than those given by energetic H injection except at elevated temperatures. For carbon materials, owing to their porous nature, molecular hydrogen can penetrate deep and tap on the surface of crystallites given homogenous depth profile as shown in Sect 9.4.1 and Fig. 9.2 in Chap. 9.

7.4 Reflection, Reemission, and Retention

Figure 7.4 schematically shows time sequences of reemission flux of injected fuels from the plasma-facing wall (a) and retention in the wall (b). Except direct reflection of the incident particle, the figure is quite similar to Fig. 4.14 in Chap. 4. The reemission flux increases with time and finally comes to the steady state where the impinging particle flux balances with the reemitted flux plus the permeation flux. Generally, the permeation flux is quite small and not considered in hydrogen recycling but does matter for T safety as discussed later. Hatched area in Fig. 7.4a corresponds to the amount of the hydrogen particles (H) retained (dissolved and trapped) ($W(t)$) in PFM,

$$W(t) = \int_0^t (\varphi_0 - \varphi(t))dt, \tag{7.3}$$

where ϕ_0 and $\phi(t)$ are incident flux and reemitted flux, respectively, and plotted in Fig. 7.4b. After the termination of H injection, if there is no H trapped, all retained are subsequently released by diffusion. While if some are trapped as indicated as bold dotted lines in the figure, the reemission after the termination of the H injection decays earlier than that without the trapping and some of injected H remain in PFM.

Because energetic H injection produces new defects in PFM, which can trap H subsequently injected, the true steady state is hardly attained. Nevertheless, the retention rate or difference between the injected flux and reemitted one becomes less and less toward the saturation. Hence, after a certain time of the injection, the remitted flux becomes constant nearly equivalent to the incident flux. However, the integrated

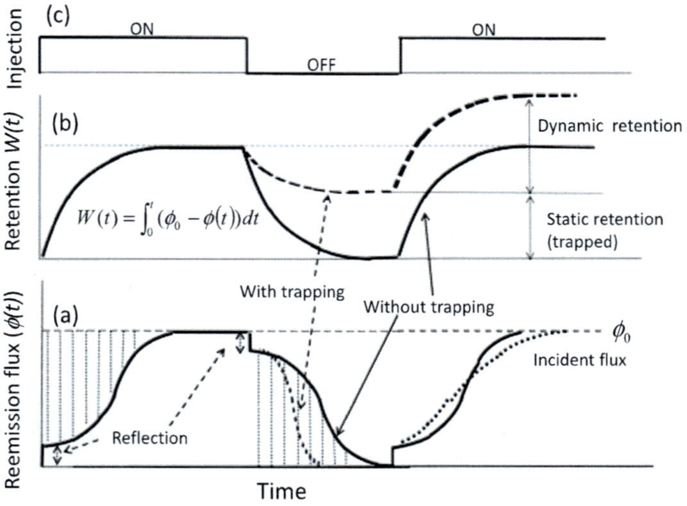

Fig. 7.4 Schematics for time sequences of **a** reemission flux of injected fuels from PFM and **b** retention in PFM. Lines and dotted lines correspond to remissions without and with trapping, respectively

amount of the retained flux gradually increases and becomes quite large after long time injection. Because H injection and neutron irradiation introduce various defects which work as H trapping sites. It is quite difficult to evaluate *W(t)* accurately after the long time exposure. In principle, the integration of the flux balance gives the *W(t)* as given by Eq. (7.3). However, the accuracy of the measurements of $\phi(t)$, only 2–3 digits, does not allow the reliable evaluation of *W(t)*.

As indicated in the figure, the total retention during the injection includes two components; "dynamic retention" and "static retention". The former is directly connected to the H recycling and the latter to the T inventory. The latter decreases with increasing the wall temperature, because trapped T can be detrapped thermally. As mentioned above, irradiation of energetic ions and neutrons produces various defects such as interstitials and vacancies, their clusters or dislocation loops, bubbles and so on, which work as additional trapping sites to increase the static retention or in-vessel T inventory, especially for W [6].

Although the permeation flux is tiny compared to the injected flux, the permeated tritium into the cooling water is a safety concern and free permeation is not allowed. Hence, formation of permeation barrier is required and is an important R&D task [7].

As discussed in Chap. 4, some of injected H are directly reflected and the flux ratio of the injected H and reflected one is defined as the reflection coefficient. Generally, the reflection coefficient is higher for heavier metals and lower incident energy. Reflected particles are either ions or neutrals, with less ions for lower energy injection. Because reflected particles do not fully loose energy, they carry energy resulting in energy reflection. Incident energy dependence of reflection coefficients of H and energy for H injection to carbon (C) and tungsten (W) [8] are shown in Figs. 4.5 and 4.6, respectively. W shows larger reflection coefficients for both particles and energy than C. The figures indicate that more than half of both incident particles and energy injected into W are reflected. This means that compared to C, plasma-wall interaction could be less on W.

It should be noted that at ITER or reactor divertor, incident H flux is expected to be 10^{23}–10^{25} particles·m^{-2}·s^{-1}, which is 10^4–10^6 times large than areal density of atoms in the materials (around 10^{18}–10^{19} atoms·m^{-2}). This means that after 1 s of injection, 10^4–10^6 atomic layers, or, 10^3–10^5 nm from the surface are fully occupied or saturated with H. If the incident energy is 100 eV or less, surface region within 10^3 nm in depth is immediately saturated. Until now we do not have such high H flux sources. There may be some new physics or chemistry. Actually, as shown in Fig. 4.12 [9], chemical sputtering of C by energetic H significantly decreases with H flux of over 10^{22} m^{-2}·s^{-1} for which reason is not clarified yet (see Chap. 4).

In permeation driven by energetic H injection, as discussed later, recombinational release of incident H is often believed to be the rate-limiting process of the reemission. For extremely high flux discussing here, recombination must be spontaneous because the surface H concentration immediately comes up to saturation and the saturation concentration must be very close to 1 in H atom to metal atom ratio (H/M). For such densely hydrogen saturated surface, the reemission of injected H is not necessarily due to molecular emission but atoms. Moreover, reflection coefficient at the H saturated layers would be modified from that of no H including surface.

In addition, at higher temperature, reemitted H is not necessary to be molecules. In chemical equilibrium in following dissociation of H_2 molecules, atomic hydrogen is preferred at higher temperatures.

$$H_2 + \Delta G = H + H, \tag{7.4}$$

where ΔG is free energy change accompanied with the reaction with temperature dependence of

$$\Delta G = \Delta H - T\Delta S, \tag{7.5}$$

where ΔH and ΔS are, respectively, enthalpy and entropy changes accompanied by the reaction. ΔG decreases with increasing temperature and is represented by the equilibrium constant as,

$$\Delta G = -RT/In\,(P(H)^2/P(H_2)), \tag{7.6}$$

where $P(H)$ *and* $P(H_2)$ are partial pressures of H and H_2, respectively. Accordingly, dissociation of hydrogen molecules become easier at higher temperatures as evidenced in Fig. 4.15, in which atomic reemissions from metals (Mo, Ta, and W) are appreciable above 1100 K under 3 keV D^{3+} irradiation at steady state [10, 11]. The atomic reemission was observed for graphite as shown in Fig. 7.5 [12]. Thus, irrespective of materials, the atomic reemission becomes appreciable above around 1000 K. The atomic reemission on the graphite limiter surface was confirmed in TEXTOR plasma [13]. This means more than half of reemitted H should be atoms on the divertor surface of ITER and a reactor of which temperature would be more than 1200 K. Hence, H recycling in a reactor would be somewhat different from that observed in present tokamak operated below 1000 K.

It should be also noted that the saturation concentration attained under energetic H injection cannot be kept after stopping the injection and major part of H in surface saturated layers must be subsequently released out. The releasing rate and total release (or how much remains) significantly change with materials and temperature. At lower fluence, there is one experiment in which release of deuterium (D) injected in Ni was measured changing its temperature as shown in Fig. 7.6 [14]. In the figure are compared the amount of D retained in Ni under the injection (referred to as injection), D released during the decay period of 3 min (evolution), and desorbed thermally (desorption) for 30 keV D^+ injection into Ni. The evolution corresponds to dissolved D or dynamic retention, and the desorption corresponds to trapped D or static retention. Above 800 K almost all D retained during the injection was immediately (within a few seconds) released after stopping the injection. Although both dissolved and trapped D decrease with increasing the temperature, the latter decreases more significantly and disappears at elevated temperatures, i.e. the static retention significantly decreases at elevated temperatures.

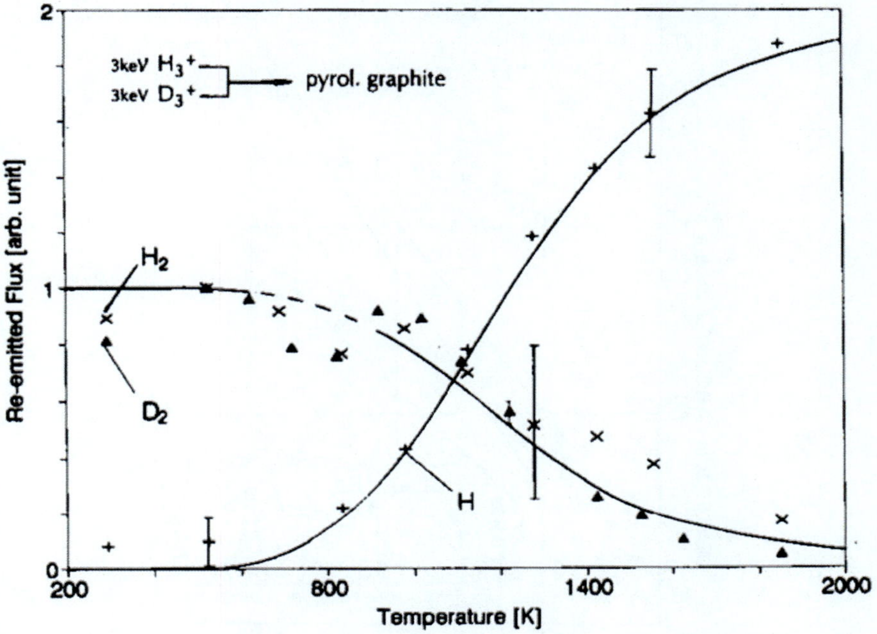

Fig. 7.5 Temperature dependence of reemission of atomic H, molecular H_2 and D_2 during 3 keV H_3^+ (D_3^+) bombardment of pyrolytic graphite measured by a line-of-sight mass spectrometer. The H_2 and D_2 reemission fluxes have been normalized at 525 K. The atomic H flux values are relative to the H_2 flux. The error bars correspond to twice the standard deviation for 14 experimental runs, each taken over 20 s. The solid lines represent the result of the model calculation (reprinted with permission from [12])

7.5 Permeation

Once a metallic membrane is exposed to H_2, H dissolved in the front surface permeate the membrane to the back surface owing to the gradient of chemical potential or concentration of retained H from the front surface to the back surface as described in Chap. 4. H permeation also occurs when the front surface is subjected to energetic H ions or plasma exposure. Abbreviation of IDP, PDP, and GDP is used for ion driven permeation, plasma driven permeation, and gas driven permeation, respectively.

Figure 7.7 shows time sequences of D permeation rate observed in typical IDP [15]. After starting D ion injection, IDP rate increases. However, it turns to gradual decay referred to as a permeation spike. This decaying tendency did not change for short interruption of the ion injection. While after a longer interruption, the permeation recovers to the initial stage and gradual decay follows again. The cause of this reduction is twofold, enhancement of either surface recombination and/or of diffusional release of injected hydrogen. Reemitted H would remove surface contaminants like oxygen and carbon to make surface clean which increases the recombination coefficient of the injected surface. This is most likely the cause of the permeation

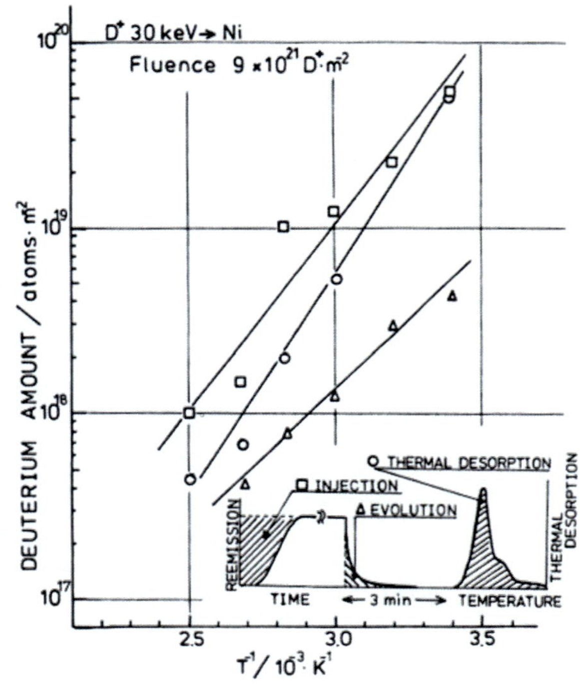

Fig. 7.6 Comparison of the deuterium amount retained in Ni under the irradiation of 30 keV D^+ ions. □: D retained during the irradiation (injection), △: release after the stopping the irradiation within 3 min (evolution or dynamic retention) and ○: D thermally desorbed after the irradiation (desorption or static retention). The lines are for the guide to the eye [14]

spike, and it repeats by contamination during the injection off and cleaning during the injection on. Since Yamawaki et al. [16] have clearly indicated that surface cleaning by Ar ion sputtering decreased IDP. In case that the surface recombination is not the rate-limiting process of H release, surface cleaning could enhance hydrogen release, for example, reduction of surface potential barrier for desorption. Another explanation is enhancement of H diffusion to the front surface from H injected region, a little deeper depth. During the injection, the defect density is increased resulting in dislocation loops. The loops tangle and hydrogen trapped at the loops easily move through the tangled loops to enhance their diffusion or mobility. When the injection is stopped, some of the defects might be annealed to recover or disappear. In either case, reduction of concentration of dissolved hydrogen in the injected region is the cause of the reduction of permeation rate.

Under extremely high flux, like ITER divertor, it is still unknown what is the main control factor to determine the surface saturation concentration. Hence, hydrogen retention and permeation in the materials under high flux and high temperature are quite difficult to estimate.

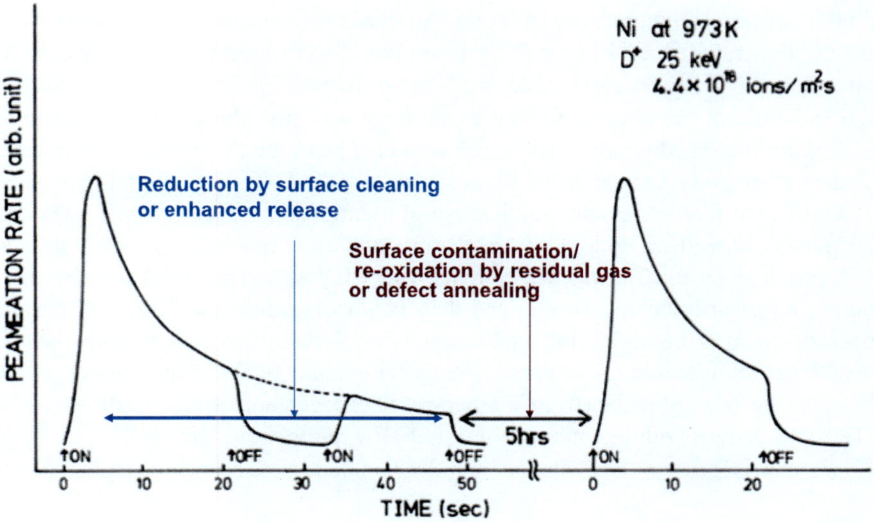

Fig. 7.7 Typical time sequences of D permeation rate for ion driven permeation through Ni under irradiation of 25 keV D^+ ion [15]. Permeation spike appeared probably owing to surface cleaning and repeated by contamination during the injection off

7.6 Isotope Effects

Because of large mass differences among H, D, and T, difference of their behavior in interactions with materials are appreciable and referred to as isotopic effects. As an example, appreciable difference was observed in blister formations on irradiated metals surface with H and D ions [17] owing to their different damaging rate. (See Fig. 6.10). The isotope effects are appreciable in plasma confinement showing higher plasma performance in D plasma than H plasma in all tokamaks making H discharge and D discharge separately. It is owing to heavier mass of D than H, resulting in better confinement time and fueling efficiency, lower escaping flux from plasma, and larger evacuation rate from the vessel for D [18]. As described in Chap. 2, the simple kinetic theory suggests the dependences of the square root of the mass ratio of D/T on those properties, but more or less no quantitative data for the isotopic effects for D and T in plasmas and in materials are available until now.

Due to the isotopic effects, it must not be easy to keep appropriate D/T ratio in plasma to continue the most efficient burning or to keep DT burning efficiency the highest without independent fueling of D and T. To do this, concentrations of D and T in the burning plasma must be separately measured to make feedback to fueling possible. It is, however, quite difficult to evaluate or analyze the amount of D and T separately in the burning plasma. Different confinement times of D and T in plasma would result in their inhomogeneous distribution particularly in radial distribution (see Chap. 8).

Furthermore, retention rates of D and T in PFM are not necessarily the same, so as their release rates from PFM are. This means that if some local power load like ELM hits the wall, thermally released D and T are different in their amounts and disturb the D/T ratio of the plasma. Owing to the huge wall inventory, a small change in its D/T ratio in PFM would cause significant change in the D/T ratio of the plasma. Thus, we are going to face difficulty in controlling the D/T ratio in burning plasma.

Until now few systematic studies using intentionally mixed hydrogen (H and D) plasma have been done. Only in JT-60U and JET, retentions of all hydrogen isotopes of H, D, and T have been systematically studied and proved that all isotopes do not necessarily behave similar and their behaviors reflect the history of plasma operations. In D operating tokamaks, a few % of H always remained in their D discharges after intense D discharges [19, 20]. For study of D plasmas, effects of the few% H are negligible, but from T aspects it is unacceptable. In the initial phase of ITER, discharges with intentional control of H/D ratios should be made.

7.7 Long-Term Retention and Trapping

In Sect. 7.4, H retention in PFM is discussed from the fuel recycling point of view. Since injected H flux is so large, the retention rate relative to the injected flux would be only a few % or less. In addition, most of retained H would be thermally released after the termination of the injection. Therefore, the increasing rate of the static inventory with discharge times would not large. Nevertheless, after long time exposure to plasma, the static inventory would pile up to be far larger than H in whole plasma.

Long-term H retention concerns on density control of burning plasma and T safety. A little temperature up of PFM would result in significant amount of T release and make density control of burning plasma difficult, because the trapping decreases with temperature, which is discussed more detail in Chap. 8.

Here considered is the T safety. Because of T regulation, any small amount of T remaining in a system must be accounted. When T inventory in any system becomes over the regulated amount, for example, in ITER it is 1 kg according to the French regulation [21], the system is not allowed to continue the operation and forced to remove T. The T inventory in a reactor, in general, corresponds to the static inventory discussed in Sect. 7.4. In addition, the initial inventory which would be significantly large should be taken into account. That is because that any materials contain H as an impurity and surface absorbed H_2O, which can be replaced by D and T injected by exposure to burning plasma. For a simple estimation, assuming that 10 ppm of H is included in stainless steel (SS) as a structure material, which is not the overestimation, and they were isotopically exchanged by D and T, the D+T retention in a whole reactor vessel consisting of 1000 tons of stainless steels could be easily over few kg [18]. This is significantly larger than the allowable T inventory in ITER. Of course, the isotopic replacement must not be spontaneous, but would proceed very slowly. This kind of inventory referred to as the initial inventory might not be distinguished from the static inventory. Thus, the static inventory gradually increases owing to diffusive penetration of fuels to make isotopic replacement with impurity H and to be trapped at defects produced by energetic particle injection including neutron.

As discussed in Chap. 3, quantitative analysis of the static T inventory in PFM is quite hard. By integrating the difference of the T amounts between throughput and exhaust, the dynamic T inventory of a system can be estimated. However, the accuracy of the T measurement in gas or liquid is limited within 2 or 3 digits. This means that the loss of 1 g T from a T handling system can hardly be detected when 1 kg of T is handled in it. Basically, the static retention can be determined by thermal desorption after the discharge [22, 23]. However, very high temperature is required for full recovery of T from PFM. T recovery is also required because T resource is not enough, which is described in Chap. 8.

Here, it should be pointed out that T inventory described here is limited to that in PFM. However, T inventory in deposits of eroded materials on plasma shadowed area could be much larger, in particular, for C-based PFM, as indicated in Fig. 7.8 [24] and discussed in Chap. 5 (see Fig. 5.6).

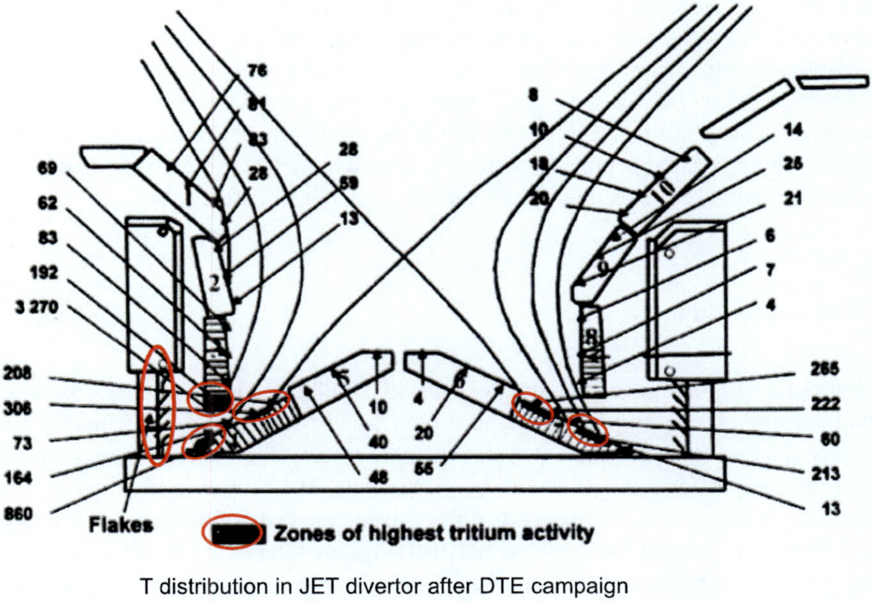

Fig. 7.8 Tritium concentrations in JET divertor tiles used in JET DT campaign. T was analyzed by full combustion (in MBq·cm^{-2}). The analysis was performed on disks (7.8 mm diameter, 1 mm thick) from the plasma-exposed side of the tiles. Tiles 1–10 correspond to JET tiles IN1S1, IN2S1, IN3S1, IBN4, IBN5, IBN6, IBN7, 10BN8, 10BN9, and 10BN10, respectively. Circled areas showing higher T retention are located at plasma shadowed area except divertor target areas. (reprinted with permission from [24]) Compare with Fig. 5.6 which shows drilled holes and surface profiles of T on the tiles

7.8 Simulation and Modeling

Numerical solution of diffusion equation using simulation codes as TMAP 7 [25] has been often used for estimation of T permeation and retention in fusion reactor materials. To simulate stoichiometry change or concentration changes of implanted ions for higher fluences, ACAT-DIFFUSE Code is developed [26, 27]. In such codes, damaging processes or modification of material properties like solubility and diffusivity given by energetic hydrogen injection are not considered. In other words, in order to simulate or analyze interactions of energetic hydrogen with materials, both process of energy transport and particle transport should be simultaneously handled. In this respect particle simulations like molecular dynamic simulations are encouraged, which are not described in this book. However, development of adequate potential to handle atoms and molecules in excited state resulting from injected energetic ions might not be enough. That means the situation is the same with experimental investigation which requires consideration of energy dissipation processes including the formation of excited atoms and molecules as discussed in this book. Hartree–Fock molecular dynamics in which potential energy is recalculated in every time step based on the Hartree–Fock molecular orbital theory seem suitable. Still, for discussion of dynamic motion of hydrogen, adiabatic approximation or Born–Oppenheimer approximation is usually taken into account in Hartree–Fock molecular orbital calculations may not be correct, because vibrational frequency of hydrogen atoms, around 10^{13} is too fast to be approximated being quiescent. For understanding of interactions of hydrogen with fusion reactor materials, theoretical modeling and adequate simulations enable to handle them are awaited.

7.9 Summary

Hydrogen (fuel) recycling (fueling, exhaust and wall retention) is the key to maintain burning plasma. Furthermore, radioactivity and poor resource of T fuel require precise control of T flow and T inventory in a reactor. Under steady-state burning, fuel throughput and exhaust should be mostly balanced. However small amount of fuel is continuously lost in deposits on plasma shadowed and remote areas and in PFM inside in the reactor vessel. Since the longtime integration of the loss, which becomes T inventory in a reactor, could be easily over-regulated T level, reduction of the T inventory both during and after the plasma operation are mandatory.

As described in Chap. 6, heating and modification (damaging) of PFM by power load and hydrogen recycling synergistically occur, understanding and controlling of hydrogen recycling is quite hard. In addition, no experimental apparatus which gives similar power load given by energetic photons and hydrogen in a reactor are available.

Hydrogen (H) behavior in metals has been long term research subjects for its importance of hydrogen embrittlement. Nevertheless, it remains as unsolved subject.

In a fusion reactor, H injected in PFM with energy gives additional processes. They appear as, reflection, damage formation including blistering, reemission of atoms and molecules, diffusion and trapping and influence of damages on them like radiation enhanced diffusion, super-permeation, and so on. In addition, mixture of D and T as fusion fuel brings isotope effects in these processes, owing to their large mass difference, which is not easy to study because handling such large amount of T used in a reactor or even in ITER is impossible in any laboratories at present. Therefore, results obtained for H and D should be extrapolated to D and T.

In this chapter, above mentioned fundamental processes are introduced with a little detail, for foreseeing or understanding what the hydrogen recycling in a reactor likely to be. Although simulation would help this, the processes are so synergistic that the present simulation is hard to cover all important processes.

References

1. M. Shimada, R.A. Pitts, Wall conditioning on ITER. J. Nucl. Mater. **415**, S1013–S1016 (2011)
2. T. Tanabe, *Interactions of Hydrogen with Fusion Reactor Materials*, National Research Nuclear (University MEPhI, Moscow, 2019). ISBN 978-5-7262-2632-3
3. T. Nakao, Chapter 4, Burning Plasma, in *Tritium: Fuel for Fusion Reactors*, ed. by T. Tanabe (Springer Japan, 2017), pp. 73–92, ISBN 978-4-431-56458-4
4. A. Kallenbach, M. Bernert, R. Dux et al., Impurity seeding for tokamak power exhaust: from present devices via ITER to DEMO. Plasma Phys. Control. Fusion **55**, 124041 (2013)
5. T. Tanabe, Tritium fuel cycle in ITER and DEMO: issues in handling large amount of fuel. J. Nucl. Mater. **438**, S19–S26 (2013)
6. T. Tanabe, Review of hydrogen retention in tungsten, Phys. Scr. **T159**(12), 014044 (2014)
7. T. Tanabe, Surface barrier for tritium permeation. Fusion Technol. **28**, 1278–1283 (1995)
8. T. Tabata, R. Ito, Y. Itikawa, et al., Data on the backscattering coefficient of light ions from solids, Rep. IPPJ-AM-18 (1981), (R. Ito, T. Tabata, T. Itoh, Data ow the backscattering coefficients of light ions from solids (A Revision) Rep. IPPJ AM-41, 1985)
9. J. Roth, A. Kirschner, W. Bohmeyer et al., Flux dependence of carbon erosion and implication for ITER. J. Nucl. Mater. **337–339**, 970 (2005)
10. J.W. Davis, A A. Haasz, Reemission of deuterium atoms from Mo, Ta and W during D^+-irradiation, J. Nucl. Mater. **223**, 312–315 (1995)
11. J.W. Davis, A.A. Haasz, Reemission and thermal desorption of D°, D2 and CD4 from graphite. J. Nucl. Mater. **220–222**, 832–835 (1995)
12. P. Franzen, E. Vietzke, A.A. Haasz et al., Atomic re-emission of hydrogen from pure and boronized graphites at temperatures above 1000 K. J. Nucl. Mater. **196–198**, 967–971 (1992)
13. A. Pospieszczyk, P. Mertens, G. Sergienko et al., In situ measurement and modeling of hydrogen recycling and transport processes–the role of molecules. J. Nucl. Mater. **266–269**, 138–145 (1999)
14. T. Tanabe, H. Hirano, S. Imoto, Re-emission and thermal desorption of deuterium implanted in Ni near room temperature, J. Nucl. Mater. **151**, 38–47 (1987)
15. T. Tanabe, Y. Furuyama, S. Imoto, Reemission and permeation of deuterium implanted into metals. J. Nucl. Mater. **122&123**, 1563–1567 (1984)
16. M. Yamawaki, N. Chitose, V. Bandurko et al., Impact of surface phenomena in metals on hydrogen isotope permeation. Fusion Eng. Design **28**, 125–130 (1995)
17. N. Enomoto, S. Muto, T. Tanabe et al., Grazing-incidence electron microscopy of surface blisters in single-and polycrystalline tungsten formed by H+, D+ and He+ irradiation. J. Nucl. Mater. **385**, 606–614 (2009)

18. T. Tanabe, Tritium handling issues in fusion reactor materials. J. Nucl. Mater. **417**, 545–550 (2011)
19. S. Higashijima, The JT 60 Team, Wall conditioning to control Impurity and hydrogen isotope recycling in JT-60U. J. Plasma Fusion Res. **75**, 1297 (1999). (in Japanese)
20. M. Yoshida, T. Tanabe, A. Adachi et al., Carbon transport and fuel retention in JT-60U with high temperature operation based on postmortem analysis. J. Nucl. Mater. **438**, S1261–S1265 (2013)
21. D. Murdoch, M. Glugla, T. Hayashi et al., Evolution of ITER Tritium Confinement strategy and adaptation to Cadarache site conditions and French regulatory requirements. Fusion Eng. Design **83**, 1355–1358 (2008)
22. J. Roth, E. Tsitrone, A. Loarte et al., Recent analysis of key plasma wall interactions issues for ITER. J. Nucl. Mater. **390–391**, 1 (2009)
23. E. Tsitrone, B. Pegourie, Y. Marande et al., Multi machine scaling of fuel retention in 4 carbon dominated tokamaks. J. Nucl. Mater. **415**, s735 (2011)
24. R.-D. Penzhorn, J.P. Coad, N. Bekris et al., Tritium in plasma facing components. Fusion Eng. and Design **56–57**, 105–116 (2001)
25. G.R. Longhurst, J. Ambrosek, Verification and validation of the tritium transport code TMAP7. Fusion Sci. Technol. **48**, 468–471 (2005)
26. S.T. Nakagawa, Y. Yamamura, Depth profiling and stoichiometric changes due to high-fluence ion bombardments, Nucl. Instr. Methods, **B33**, 780–783 (1988)
27. T. Ono, T. Kenmotsu, T. Muramoto, T. Kawamura, Calculation of deuterium retention, re-emission and reflection from a tungsten material under D^+ ions irradiation with ACAT-DIFFUSE code. J. Nucl. Mater. **390–391**, 713–716 (2009)

Part III
PMI, Observations in Present Large Tokamaks and Prospects in a Reactor

Chapter 8
PMI in Large Tokamaks

8.1 Power Load

As shown in Chap. 2, research targets have been changing with improvement of plasma confinement. Now, targeting the establishment of a fusion reactor as an energy source, durability of plasma-facing materials (PFM) to steady and transient power load are main concerns. Nevertheless, no tokamaks could have given such high-power load as expected in a reactor. Only JET is now working to give power load a little less than that expected in the reactor. However, after changing its plasma-facing wall from carbon-based materials to ITER like Wall, i.e. Be for first wall and W for divertor, the power loads are carefully controlled not to destroy the W divertor. In this respect, the previous carbon-based divertor allowed higher power load and PMI data including effects of giant ELM and disruption are used for design base for ITER.

Recent large power linear plasma machines are also useful to examine material responses to the high-power loads as already noted in Chap. 2 and in reviews [1, 2]. However, owing to their different geometrical structure from divertor and lack of strong magnetic field, it is questionable to use them as a divertor simulator except power load tests. In this respect, data obtained in presently running plasma machines under well-designed program for studying plasma material interactions (PMI) would be useful. In this chapter, PMI observed in large tokamaks, in particular, JET and JT-60U are mainly summarized.

8.1.1 *Power Load in JET*

To ensure sufficient divertor target lifetime, the loss of plasma stored energy due to ELMs in ITER should be restricted to $\Delta W_{ELM} \leqq 1$ MJ. Only in JET, by virtue of its size, can such energies be approached. And the impact of large type I ELMs

T. Tanabe, *Plasma-Material Interactions in a Controlled Fusion Reactor*, Springer Series in Plasma Science and Technology, https://doi.org/10.1007/978-981-16-0328-0_8

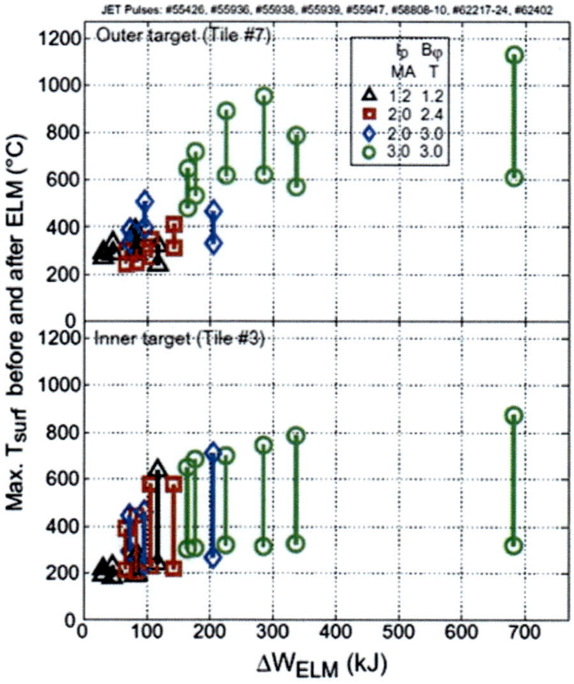

Fig. 8.1 Peak surface temperature of divertor (T_{surf}) before and during ELMs in JET. Results are obtained from coherent averaging of ELM groups near the end of the H-mode phase (Reprinted with permission from [3])

in high current H-mode JET discharges with ITER-relevant pedestal characteristics have been examined. As seen in Fig. 8.1, the increase of nearly 600 °C was observed in peak divertor temperatures during ELM in JET [3].

Recent results on ELMs in JET are summarized in [4] as follows:

(1) The ELMs provoke strong radiation losses, mostly confined to the inner divertor region. Large type I ELMs with $\Delta W_{ELM} \geqq 0.72$ MJ show enhanced radiation losses which are associated with the ablation of carbon layers in the inner divertor. Such large ELMs are usually followed by a phase of type III ELMs with an increased radiation in the plasma core.

(2) The unmitigated disruptions exhibit small radiation fractions with strong poloidal radiation asymmetry during the thermal quench, which could cause Be melting by radiation in ITER. In dedicated experiments on massive gas injection, about 60% of the thermal energy and a significant part of the magnetic energy was converted into radiation and spread uniformly over the first walls.

It is also important to note that even during ELM in JET with ITER like wall, W source and W content in plasma are separated as shown Fig. 8.2 [5]. Under normal discharges, the separation of impurity source and its accumulation in plasma was already noted in high Z test limiter experiments in TEXTOR as shown in Chap. 2 (see Fig. 2.11).

Disruptions in tokamaks lead to significantly large power loads onto the limited/localized area of plasma-facing components (PFC) and various efforts have

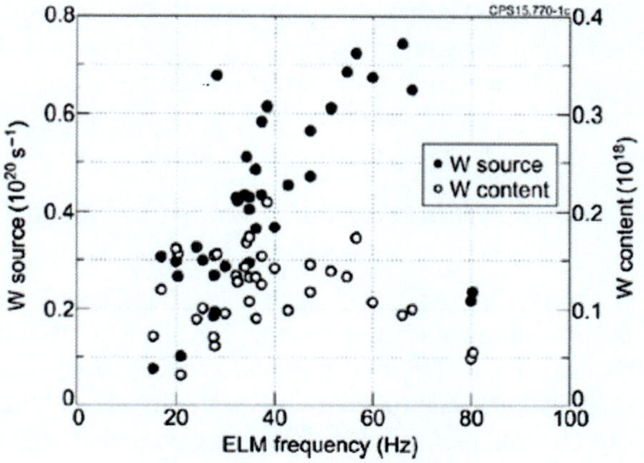

Fig. 8.2 The total W content of the plasma and the outer divertor tungsten source as function of the ELM frequency in JET. Despite the large scatter, above an ELM frequency of about 40 Hz, the tungsten content of the plasma stabilizes while the source still increases. This can be interpreted as a sign of ELM flushing (Reprinted with permission from [5])

been done for mitigation of the disruption. Nevertheless, it seems quite hard to fully avoid the disruption. Two processes, of particular concern for the first wall integrity, have been studied in dedicated experiments at JET [6]: (1) During the thermal quench, it is measured (using fast IR thermography) that 5% of the plasma stored energy is deposited onto the outer and inner poloidal limiters. More surprisingly, during the current quench, about 10% of the magnetic energy is deposited onto the outer and inner poloidal limiters via plasma surface interactions (PSI). (2) Very localized heat loads due to runaway electrons (RE), generated in disruptions triggered by massive injection of argon and neon, are measured onto the JET upper dump plate. Figure 8.3 shows averaged temperature increase measured on the JET upper dump plate due to runaway electrons (RE) impact as a function of its current [6]. Above 400 kA of the RE current, the temperature increase becomes appreciable.

To reduce power load to PFS, discharges with impurity seeding like N_2, Ne, and Ar have been done. Significant changes are reported, not only in the physics directly

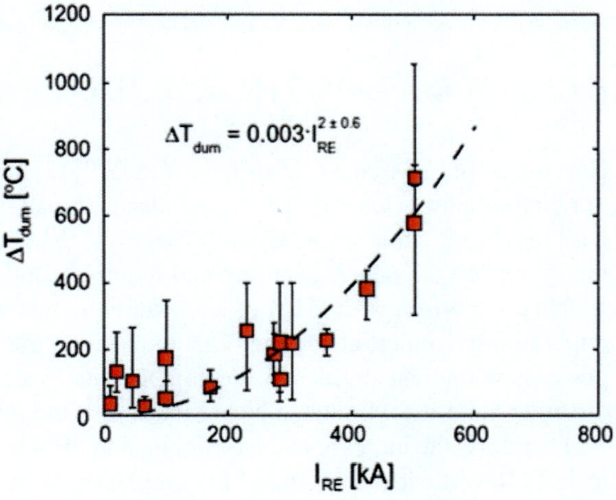

Fig. 8.3 Averaged temperature increase measured on the JET upper dump plate due to runaway electron (RE) impact as a function of the RE current. Error bars indicate the minimum and maximum temperature increase. The dashed line indicates the fit (Reprinted with permission from [6])

related to PMI but also to the main plasma which is strongly affected in unexpected ways, impacting many aspects of tokamak operation. In addition, the increase of the erosion by the seeded impurities is also concerned.

8.1.2 Exchange of PFM from Carbon to High Z Metals

As introduced in Chap. 1, high Z PFM had been avoided until recently because of the fear on high radiation from high Z elements accumulated in plasma center. After finding that source, transport and accumulation of high Z elements in plasma are separated and controlled by different physics as observed in TEXTOR Mo limiter experiments [7], and that central accumulation could be avoided with appropriate plasma control, several plasma machines introduced high Z materials as PFM. For examples; ASDEX-U gradually increased surface coverage of its PFM with W [8], Alcator C-mod operated with a first wall and divertor tiles made entirely of Mo [9], and EAST has changed PFM from initially introduced full C to W coated C [10]. Now, JET is under operation with ITER-Like Wall (ILW) consisting of Be as first wall and W coated C as divertor. The first summary on comparison of low Z and high Z used as PFM appeared in [11]. Low Z and high Z elements respond to plasma quite differently, as described in Chap. 4, and conversion of PFM from low Z to high Z has resulted in significant change in plasma behavior. Unfortunately, however, owing to concerns on melt damages of high Z materials, power load to PFM, particularly, limiter and divertor, was limited, which made it difficult to investigate responses of high Z elements to reactor like power load. Therefore, high-power load tests have been done mostly linear plasma machines or high-power ion and electron guns (see Chap. 3).

Nevertheless, JET ILW is giving important and useful information on possible PMI in a reactor. In the following is introduced a little detail of different observations between Carbon (C-wall) wall and ILW in JET.

8.1.3 ITER-Like Wall (ILW) in JET

Conversion of PFM from Carbon to ILW in JET [12] has resulted in significant change of radiation power load to PFS during plasma break down as shown in Fig. 8.4 [13] with higher radiation in the C-wall. However, the absence of active cooling forces limitations on the power handling and requires a much more cautious operation in JET in comparison with ITER or a reactor equipped with actively cooled PFC. JET relies simply on inertial cooling. This requires on the one hand maximization of the power-handling capabilities of the components by design optimization [14] and on the other hand strict limits on power loads to avoid melting [15].

Therefore, the increase of the power load to PFM is not the main object of the ILW in JET. Nevertheless, the main PMI processes: (a) material erosion and migration,

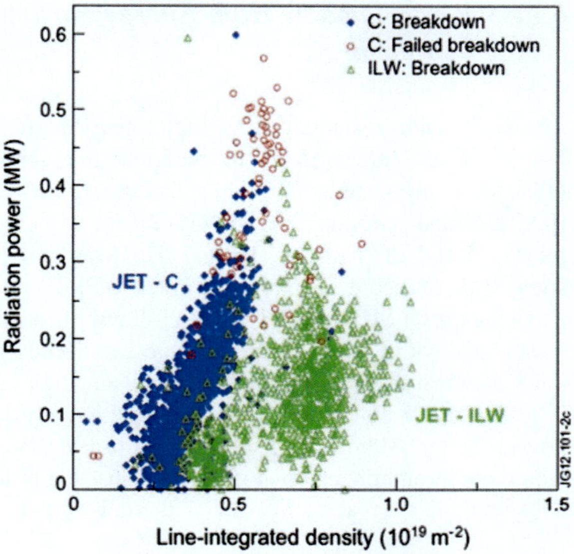

Fig. 8.4 Radiation versus density at the end of the burn-through phase in JET (Reprinted with permission from [13])

(b) fuel recycling and retention, (c) impurity concentration and radiation have been studied and compared between JET-C and JET-ILW. It is revealed that the influence of PFM on plasm performance is quite large. Brezinsek et al. [16] have summarized main observations as follows:

(a) In the H-mode plasmas for JET-ILW, primary erosion source is low and the material migration from the main chamber to the divertor is lowered by a factor of 7 and that from plasma facing to remote areas within the divertor by a factor of 30–50. The energetic threshold for Be sputtering minimizes the primary erosion source and inhibits multi-step re-erosion in the divertor. The physical sputtering yield of W by Be ions is as low as 10^{-5}.
(b) The long-term fuel retention in JET-ILW is reduced by factor 10–20 compared with that in JET C-wall. The remaining retention is caused by implantation and co-deposition with beryllium and residual impurities. Outgassing has gained importance and impacts on the recycling properties of beryllium and tungsten (Hydrogen retention is discussed in Sect. 8.4).
(c) The low effective plasma charge ($Z_{eff} = 1 \sim 2$) and low radiation capability of beryllium reveal the bare deuterium plasma physics. Moderate nitrogen seeding, reaching $Z_{eff} = 1.6$, restores in particular the confinement and the L-H threshold behavior. ITER compatible divertor conditions with stable semi-detachment were obtained owing to a higher density limit with ILW. Overall JET demonstrated successful plasma operation in the Be/W material combination and confirms its advantageous PMI behavior and gives strong support to the ITER material selection.

8.1.4 *Power Load by High Energy Particles Produced by Fusion*

In a D-T fusion reactor, various high energy particles are generated by D-T and D-D reactions. Alfa particles (He with maximum energy of 3.5 MeV) and neutrons (14.1 MeV) are generated in major D-T reactions. Additionally, T (3.5 MeV), ^{3}He (0.82 eV), and neutrons (2.45 MeV) are generated in minor D-D reactions. Although most of T and alfa particles (He and ^{3}He) lose their energy to heat burning plasma, some of them escape from plasma, inject to PFS, and deposit their energy.

Using the OFMC code [17], which follows fusion-produced charged particles and is validated with JT-60U experiments, loss of the energetic particles from the plasma due to ripple loss of TF (toroidal field) coil and their injection to PFM was calculated [18]. Figure 8.5 shows alpha particle deposition profiles to the wall in a 2 GW fusion power reactor. Although, the power load to PFM caused by the energetic particles would not be significant, the loss would cause the following issues related to PMI in a tokamak power reactor with 2 GW thermal output.

(1) If the present allowance for the ripple loss is applied, the alpha particle flux to PFS is expected to be as high as 2×10^{18} m^{-2}·s^{-1} with average impact energy of 1 MeV. When the first wall is made of W, the erosion rate by the physical sputtering of the alpha particles becomes ~20 μm per burn-year. Furthermore, blister formation and blister erosion of metal wall are concerned.

(2) Production of energetic tritons and protons is smaller than the alpha particles by two orders of magnitude. It must be noted that the deposition region of them on the wall is the same as that of alpha particles. However, the tritons would raise T inventory in PFM, because they are directly injected in PFM and continuously piled up which is discussed in Sect. 8.4.

8.2 Erosion and Deposition

8.2.1 *Carbon Wall (C-Wall)*

As noted in Chap. 5 (see Fig. 5.1), erosion and deposition occur at different locations with each other resulting in materials transport in a reactor. Generally, inner first wall is eroded, outer first wall is eroded or deposited depending on plasma character, out divertor is erosion dominated and inner divertor is deposition dominated. Owing to very high-power load by radiation and particle fluxes, the temperature of PFS is raised. In particular, the temperature rises of divertor target area (or foot prints) are largest and appreciably influence erosion/deposition rates and thermal properties of PFS.

Erosion/deposition properties are different between C-wall and metallic wall, especially W-wall. Nevertheless, general pictures of material transport observed in tokamaks with the C-wall (see Fig. 5.1) is valid in JET-ILW as discussed in the next

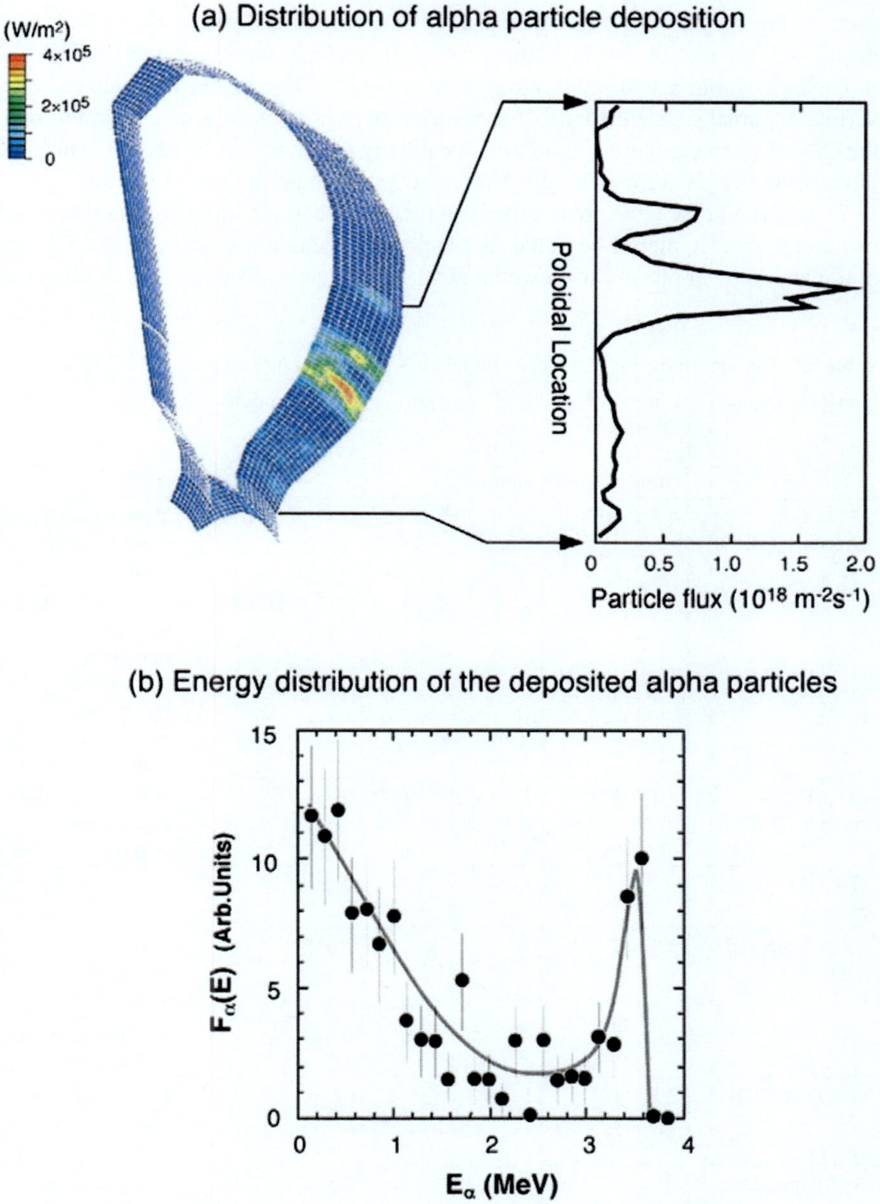

Fig. 8.5 Alpha particle deposition to the wall in a 2 GW fusion power reactor; **a** the power flux and the particle flux of alpha particles, and **b** energy distribution of the alpha particles deposited on the wall. Error bars are attributed to the statistical error in the Monte Carlo calculation (reprinted with permission from [18])

section. For the C-wall, large database on erosion and deposition is available and details of local erosion and deposition in C-wall machines were described in Sect. 5.2 in Chap. 5. Table 8.1 [19–22] summarizes overall C deposition rates observed in various tokamaks. Interestingly, it is revealed that most of tokamaks operated with the C-wall show similar erosion rates. As already noted in Figs. 5.1 and 5.2, most of the erosion occurs at the outer divertor and deposition on the inner divertor.

Data given in the table are not likely to include C deposition in plasma shadowed or remote area. It should be noted that appreciable amount of depositions appears in tile gaps, and at plasma shadowed and far remote areas. Deposition was observed

Table 8.1 Carbon deposition and fuel retention rates in current large tokamaks [19–22]

Device/Campaign	Carbon deposition rate (C atoms·s^{-1})	Fuel retention rate ((D/T) atoms·s^{-1})	Fuel retention fraction	Remarks
JETMK-IIA divertor (DTE)	6.5×10^{20}	5.8×10^{20} (D/C = 0.8)	0.17(in DTE1)	– T retention after non-mechanical cleaning
			0.11(in DTE2)	– T retention after long term outgassing and mechanical removal of accessible T deposit
JETMK-IIB divertor (DTE)	4.3×10^{20}	1.25×10^{20} (D/C = 0.3)	0.03	– D retention from post-mortem analysis
TFTRDT campaign			0.16	
ASDEX-upgrade (DD)	3.5×10^{20}		0.035	– D retention from post-mortem analysis
			0.1	– D retention from fuel balance
TEXTOR (DD)	2.5×10^{20}	1.6×10^{19}	0.08	– D retention from post-mortem analysis
Tore Supra (DD)		2.5×10^{20}	0.5	– D retention from fuel balance in dedicated long pulse discharges
JT-60U (DD)	$3 \sim 6 \times 10^{20}$ (only plasma facing area)	5.3×10^{18} (D/C = 0.02)	0.0 (for saturated wall)	– Wall saturation appeared at 573 K operation

DD: D discharges with D NBI heating

even in the bottom of gaps. Thus, deposition profiles on shadowed area and remote area are quite dependent on geometries of the areas so that it is hard to describe general picture of carbon transport (erosion and deposition) in tokamaks.

Since all tokamaks given in Table 8.1 except JT-60U which was operated above 500 K were auxiliary cooled, their base temperatures were ambient (below 400 K) and the temperature of PFS was raised during the discharge with a different raising rate and highest temperature with each other. Nevertheless, the difference in the erosion/deposition rates among them were within factors. The reason was not clear. The C content in the plasma of these tokamaks was also not so different from each other, a few %, which must be correlated to the deposition rate. If these erosion/deposition rates are simply piled up for 1 year, it becomes 1×10^{28} C atoms $= 1.6 \times 10^4$ mol or $= 1.9 \times 10^5$ g. That is, nearly 190 kg·y^{-1} of C is eroded/deposited. Suppose T retention rate in the deposited C is 0.1 in T/C, 9.6 kg·y^{-1} of T would be retained in the deposited C. Compared to allowable T inventory in ITER, 1 kg, this is too large. Thus, utilization of C as PFM in ITER was excluded. However, hydrogen retention in JT-60U operated at higher temperature was much less. Therefore, in a reactor with PFM made of C of which base temperature of PFM should be above 700 K for power generation, T/C would be less than 0.1. Even so, 0.96 kg·y^{-1} of T would be still too high. Hence, the removal of T from PFM is mandatory. More details of erosion/deposition and fuel retention are discussed in the following sections.

8.2.2 Metallic Wall

It is reported that the general picture of material transport in JET C-wall (CW) was still valid in JET-ILW with Be first wall and W divertor [23, 24]. Erosion and local deposition of Be from and on the limiters occurred during the limiter plasma phase. During divertor plasmas, Be was eroded from the inner wall by charge exchange neutrals and deposited at the inner divertor region. Compared to JET-CW, there was an overall reduction of the deposition in the JET-ILW, which should be accompanied by the reduction of net erosion from the inner main chamber wall.

Figure 8.6 shows profiles of deposition (b) and D retention (c) on divertor tiles of JET-ILW [25]. As depicted in the figure, the deposition profiles on the divertor tiles well correspond to D retention. The strike points showed less deposition and D retention compared to other regions owing to the temperature rise. Fuel retention is discussed separately in Chap. 9.

Table 8.2 are compared experimental results from long-term expose probe samples (LTS) installed at the first wall during 2005–2009 (JET CW), 2001–2012 (JET ILW campaign 1 (ILW1)), and 2013–2014 (ILW2) campaign [26]. In erosion of Be, chemical sputtering with the form of hydride (BeH_2 and BeD_2) is largely contributing [27]. In addition, less threshold energy and yield of sputtering of Be compared C, erosion rate of Be is a little larger than C. The erosion of W is much less, more than one order of magnitude, owing to its high threshold energy and less sputtering yield.

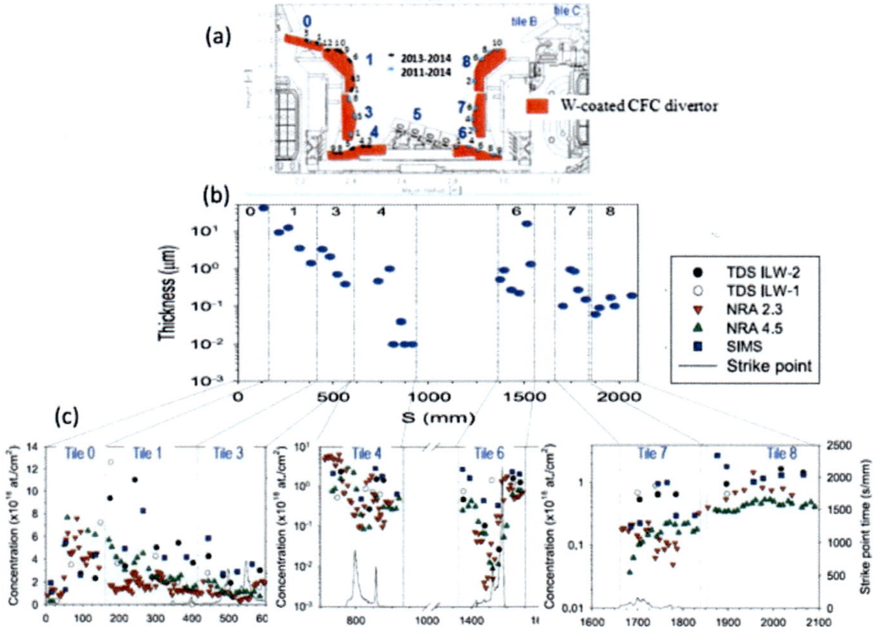

Fig. 8.6 Deposition and D concentration on divertor tiles of JET ITER like wall. **a** the geometry of the divertor, **b** thickness profiles of the deposits, and **c** Concentration of D in tiles 0–8 determined with thermal desorption (TDS), nuclear reaction analysis (NRA) and secondary ion mass spectroscopy (SIMS). TDS ILW-1 corresponds to TDS analyses made from ILW-1 tiles. NRA and SIMS results are for ILW-2. The frequency profiles of strike points on the outer divertor is also shown (reprinted with permission from [25])

Table 8.2 Comparison of average net total LTS erosion, and net erosion rates for the 2005–2009 (JET-C), 2011–2012 (ILW-1), and 2013(ILW-2) campaigns [26]

Campaign	Sample material	Average total erosion, 10^{19} atoms·m^{-2}	Average erosion rate atoms·m^{-2}
2005–2009	Be	36,000	1.2×10^{18}
	C	74,000	2.4×10^{18}
2011–2012	W	60	9.2×10^{15}
	Be	35,000	5.5×10^{17}
2013–2014	W	63	8.8×10^{15}
	Be	1,900	2.0×10^{17}

8.3 Dust

Various kind of dusts are produced in a reactor vessel, for example, exfoliation of deposited layers, fracture erosion of modified surfaces by high power loaded (recrystallized or re-solidified melt layers), droplet of the melt layers, and so on. Disruption and large ELMS very likely trigger the dust production. Figure 8.7 is an example of dust production during a disruption in JET [28].

Most of the dusts created by the discharges and disruptions are deposited at the bottom area in particular remote area as shown in Fig. 8.8 as an example [29]. Collected dusts in divertor region of JT-60U, a full carbon tokamak operated mainly at around 473–573 K, are summarized in Fig. 8.9 [30]. As seen in the figure, not only the gravity but also the flows of neutrals and fuel molecules motivate the dusts to move to ducts for divertor pumping. However, the sources of the dusts are not clear.

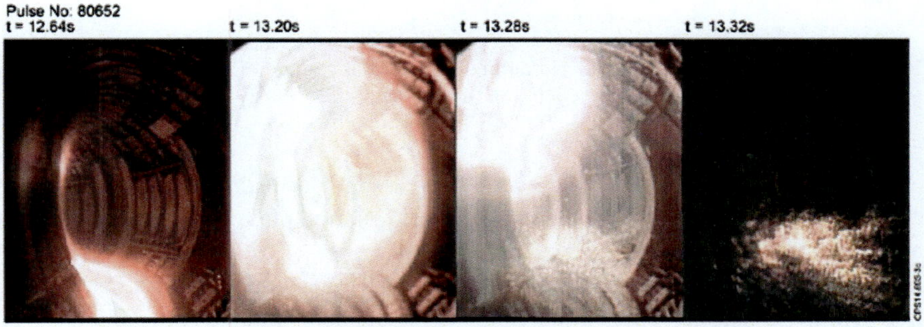

Fig. 8.7 Images from JET's COHU color CCD camera (KL1) from octant 4 for JET pulse 80652. The plasma is terminated by a disruption with $t_d = 13.14$ s and $t = 13.18$ s. Induced vessel forces are recorded as 0.43 MN. A cloud of debris is seen at $t = 13.20$ s and in subsequent frames. Note that the images in frames after $t = t_d$ show distortion due to camera shaking (reprinted with permission from [28])

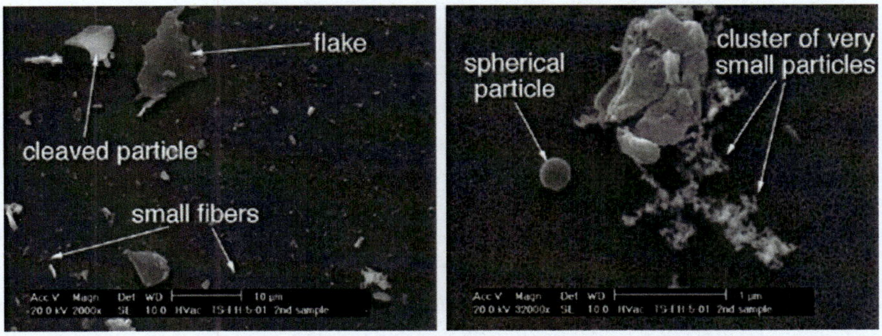

Fig. 8.8 Representative SEM micrographs demonstrating various particulate shapes of dust collected from fusion devices The shapes of some dusts suggest how they were produced but do not show where they came (Represented with permission from [29])

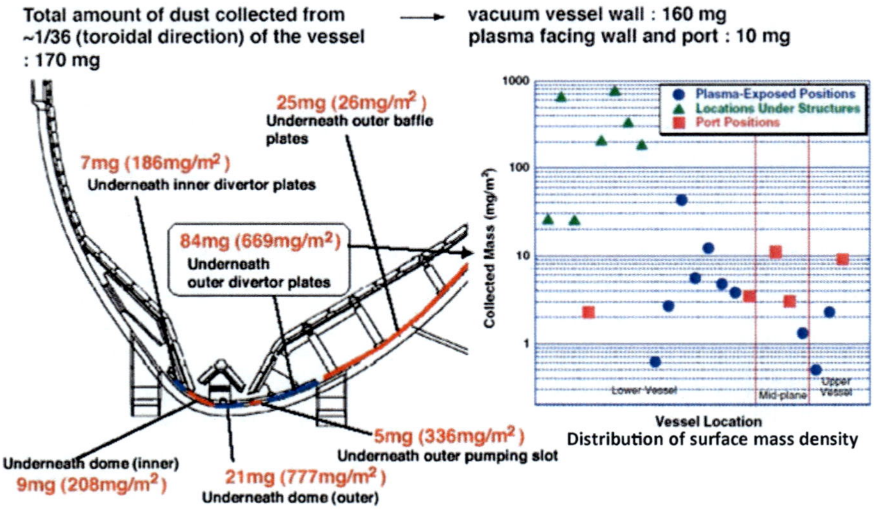

Fig. 8.9 Collected dusts in divertor region of JT-60U a full carbon tokamak operated mainly at around 473–573 K (reprinted with permission from [30])

Some are directly redeposited at the remote area with loose adhesion to the substrate owing to lower temperature, some caused by exfoliation of deposited layers on PFS and some made by disruption.

Although the amount of dust mobilized after disruptions is known to be proportional to the dynamic vessel forces, it is not possible to infer what fraction of the dust (if any) might be created by disruptions. However, disruption mitigation which is mandatory in a fusion reactor is found to reduce the amount of the dust seen after moderate disruptions by a factor of 4 [31].

Dust has not yet been of a major concern in existing tokamaks mainly because the quantity is small, and they are not nuclear facilities. Since dusts in a reactor contain tritium and neutron activated elements, they will represent operational and potential safety issues. From a safety point of view, they must be controlled based on a defense in depth approach designed to provide reliable confinement systems, to avoid potential hazard and failures, and to measure and minimize the dust inventory [31].

Dust formation critically depends on the kind of PFM. In general, higher erosion results in larger deposition and consequently more dusts. In case of C-wall, owing to its chemical erosion, significant amount of C is transported and deposited at areas facing less plasma flux and shadowed from plasma. Depending on the temperature of deposited area, the properties of the deposited layers, their fuel retention, and their micro-structure are changed. At lower temperature area, deposited C exhibits amorphous like structure with high fuel retention, while at higher temperature area, it changes to layered structure but not fully graphitized. For the case of W, the deposited layers include a lot of impurities in addition to fuel, such as oxygen, carbon and often show amorphous like structure at lower temperature area. These deposited layers are easily exfoliated resulting in dusts due to the different expansion coefficients between the layers and the substrate when large power is loaded.

Owing to less erosion rate, the deposited amount for W-wall is much less than that for C-wall. Consequently, observed dust level of JET-ILW is orders of magnitude less compared with that of the JET-CW. However, when the power load is high enough to give surface melting of the metallic wall, released droplets from the melting surface become dust. Furthermore, re-solidified layers are easily cracked to give particle ejection from the surface under high heat load, which is a major concern for utilization of W as PFM.

In a reactor, there is an additional concern. The resuspension of dust could be a consequence of lss of coolant accidents (LOCA) and lss of vacuum accidents (LOVA) and it can be dangerous because of dust radioactivity, toxicity, and capable of causing firing and/or explosion [32].

It should be mentioned that there is a lack of experimental data on thermo-chemical properties and shape of dust particles, dust production rates are unknown, and there are uncertainties in the dust–plasma and dust–wall interactions. Hence, much more studies in dust-related areas are required including improvement of dust measurement, dust cleaning, and the reduction of dust production.

8.4 Recycling and Retention of Fuels

Since the burning efficiency of fuel in a fusion reactor is poor, large amount of fuels are put through and most of them are exhausted. Moreover, fueling efficiency, which means the fraction of fuels going into plasma to the total fuel throughput is also rather poor. Therefore, any loss of the fuels in a reactor including consumed ones by burning, i.e. the difference between the throughput and the exhaust, influences fuel self-sufficiency. In addition to the fuel loss in a reactor or in-vessel T retention, some loss cannot be avoided in all T processing systems for fueling, exhaust, breeding, recovery, refinement, isotope separation, and storage. Among all, the in-vessel T retention (mainly in PFM and deposited layers) is the largest. Without establishment of the fuel self-sufficiency by T breeding in each rector, enough number of fusion reactors as energy sources cannot be built. Thus, the in-vessel T retention should be determined precisely not only for the fuel self-sufficiency but also T safety. In these respects, T breeding and recovery are also quite important engineering tasks. However, they are out of scope of this book and not discussed. Please refer textbooks such as [32]. The following sections concentrate on the in-vessel fuel retention.

8.4.1 Consideration of Fuel Retention Rate

Figure 8.10 shows simplified fuel flow based on Fig. 7.1 with normalized fuel throughput of 200 fuel particles (100 (D) + 100 (T)) per unit time under steady-state burning [33]. In the figure, the fueling efficiency, the burning efficiency, the fuel retention rate in a vacuum vessel (VV), and the T breeding ratio are set to be 0.2, 0.05, 0.02, and 1.2, respectively, as desirable values. For the total throughput of 100/100 (D/T), only 20/20 are going into the plasma and 1/1 fuels are burned to produce 1 neutron with energy of 14 MeV, and 1/1 (D/T) are retained in VV. Consequently, 98/98 (D/T) are exhausted. In present plasma apparatuses, light hydrogen (H) is always remaining as an impurity as described in Sect. 8.4.2.2. Its origin is partly residual H_2O in VV and adsorbed one on materials surfaces. Another source is back-feed from vacuum pumps. Here is assumed around 2% always remains in fuels. Whatever the sources are, additional effort to remove H in an isotope separation process is required.

Neutrons produced by the DT reactions enter blanket systems to deposit their energy to coolant for power generation and to breed T simultaneously. In Fig. 8.10, under the assumption of the breeding rate of 1.2, 1.2 T atoms are bred from 1 neutron. As a whole, 99.2 (=98 +1.2) T could be recovered for the refueling. Therefore, in this simplified case, the T fuel self-sufficiency is not attained. D can be employed from outside. Among all efficiencies in the figure, only the fuel retention rate can be changed by engineering. If the fuel retention rate could be reduced one order of magnitude to be 0.002 which is far less than the presently observed ones, the

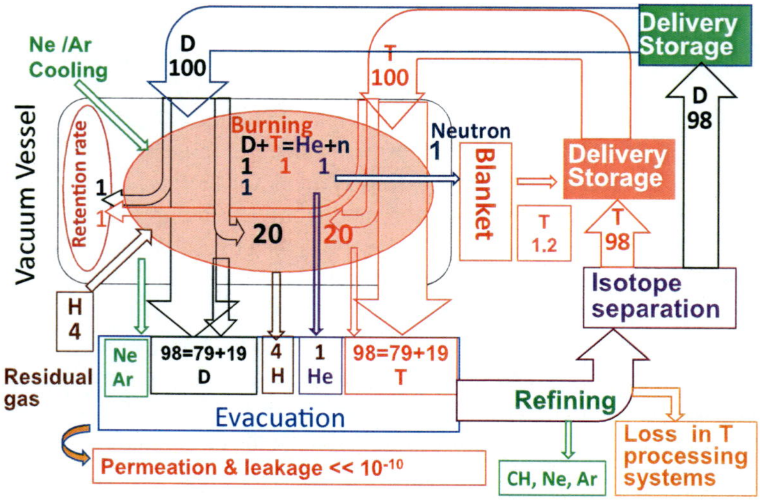

Fig. 8.10 Fuel flow and balance in a reactor at steady state burning under assumptions for overall burning efficiency: 1%, breeding ratio:1.2, fueling efficiency: 20%, and fuel retention rate: 1%. Fuel throughput of D/T is normalized to be 100/100. H always remains as a residual gas. Ne and Ar are seeded for edge cooling

recovered T became 100 (=100−1−0.2 + 1.2) and the T fuel self-sufficiency was attained. Otherwise, recovery of T retained in PFM must be done. In the present tokamaks, the fuel retention rate is as larger as a few % or more. Owing to higher temperature operation of a reactor, the fuel retention rate could be reduced. Still some T loss in other T handling systems cannot be avoided. The fuel retention rate of 0.2% is too large, while attaining 0.2% seems quite hard. Therefore, the recovery of retained T in the reactor is necessary.

In Fig. 8.10, the same fueling rates are assumed for T and D. However, behaviors of D and T would be significantly different from each other, not only in plasma but also in materials, owing to their very large mass difference. The differences would appear in confinement times in plasma, fueling efficiencies, escaping fluxes from plasma, retention rates in materials, and evacuation rates from the reactor vessel. A simple kinetic theory indicates that the square root of mass ratio of D/T could be correlated to those properties as described in Chap. 2. Nevertheless, more or less no data are available for D and T until now.

Furthermore, the different confinement times of D and T in burning plasma would result in their inhomogeneous radial distribution. Therefore, it must not be easy to keep appropriate D/T ratio in the burning plasma to get the highest burning efficiency. The concentrations of D and T in the burning plasma must be separately measured to make feedback fueling of D and T independently. It is, however, quite difficult to measure the concentrations of D and T in plasma separately and a new task for ITER [34].

The retention rates of D and T in PFM are not necessarily the same, so as their release rates from the wall are. Behavior of fuels in PFM is separately described in Chap. 5. Thus, we are going to face difficulty in controlling D/T ratio in the burning plasma to attain and keep efficient burning. However, until now no systematic efforts have been done to control two isotopes, even for H/D in plasma. All these problems remain to be solved in ITER.

In the following, detailed fuel flows in a fusion reactor are described focusing on T with special care on its radioactivity. Particularly, requirements of precise T accountancy in all T handling systems and reducing T retention in PFM as small as possible for the T fuel self-sufficiency in a fusion reactor are considered. Although handling of D/T mixture might give additional difficulty in the fuel flow control, it is not considered here.

8.4.2 Recycling

Fuel particles incident to PFS are either released back or retained in PFM. The released ones consist of two components; one reflected keeping some of their incident energy and the other reemitted mostly as molecules nearly thermalized at the temperature of PFS. Most of the reflected ones are directly return to plasma, while among the reemitted ones only ionized ones in the boundary plasma return to plasma. From plasma side, fuel particles returning to the plasma are referred to as recycled ones. And the fraction of the retuning fuel flux to the plasma compared to the incident fuel particle flux to PFS which is referred to as a recycling coefficient or rate is critically important for plasma density control.

8.4.2.1 Changes of Recycling Coefficient with Discharge Time

As described in Fig. 7.4, in early discharge time, most of the fuel particles injected to PFS is retained in PFM so that the recycling coefficient is very small. Soon reemitted fuel particles appear and increase with time. Reflection stays nearly constant except small changes caused by the fuel retention in PFS. When the fuel particle concentration in PFS layers exceeds its saturation concentration, or attained steady-state discharge, most of injected fuel particles are reemitted, i.e. the recycling coefficient becomes nearly 1.0.

Although after long discharge time or under the steady-state discharge, the recycling coefficient becomes nearly one, some small part of the incident flux continues to be retained owing to diffusion in deeper region and piled up in PFM with discharge times as fuel inventory. In principle, the inventory could be determined with the integration of flux difference between throughput and exhaust. However, the difference is not large enough to make accurate measurements. Compared to the throughput which can be determined rather accurately, the measurement of the exhaust includes large error. The fuel inventory is separately described later. Such recycling and retention scheme are basically similar for the C-wall and metal wall, although dependencies on discharge time or particle fluence are different from each other.

The retention is not limited to PFS or PFM, but that in the redeposited layers irrespective of facing or shadowed to plasma. For the C-wall, the fuel retention in deposited layers on plasma shadowed area dominates the whole retention. Because the hydrocarbons produced by the chemical sputtering enter in boundary plasma and impinge again to make deposited carbon layers. Since the deposited carbon layers are active to H, incident H onto them or residual H are retained until the layers are saturated with H. The saturation concentration in the deposited carbon layers on the plasma shadowed area is higher than those on the plasma-facing surface. Accordingly, fuel retention in the deposited layers on plasma shadowed or remote dominates in-vessel fuel inventory as discussed in the next section.

For the C-wall, the fuel particles are mostly retained making chemical bond with C atoms either C-D or C-T until the concentration attains a saturation level. The saturation concentration in C at RT is as high as ~ 0.4 in D(T)/C atom ratio. At higher temperature, it decreases significantly (see Fig. 9.3). Although the top surface is immediately saturated after plasma exposure, the saturated layers grow deeper zone with time. Hence full coverage of plasma-facing surfaces with C would result in significantly large T inventory which is one of the reasons for excluding the C-wall in ITER. However, it should be noted that since any hydrocarbons are not stable at elevated temperatures, which is the reason for the reduction of chemical sputtering above around 800 K, the saturation concentration at elevated temperatures decreases appreciably and becomes similar level to that of metals (see Fig. 9.7).

In metallic walls, the fuel particles are retained either as adsorbed water, dissolved or trapped at defects or impurities and molecules in surface blisters or babbles. Different from the C-wall for which most of the fuel inventory is caused by deposited layers, the fuel inventory in metals is caused by diffusion into deeper region showing square root dependence on plasma exposure time which is a typical diffusion-controlled retention and the retention in W obeys this. However, the retention in early times is owing to trapping at various defects in near-surface regions produced by the injected fuel particles. Later, trapping at defects produced by neutron irradiation becomes additional inventory.

In case of Be, it is known that chemical sputtering making beryllium hydride produce deposited layers which increases the fuel retention [27]. Therefore, erosion, deposition and fuel retention schemes for Be would be similar to those for C.

Irrespective of C or metals, the wall inventory is larger by several orders of magnitude than the total number of fuel particles confined in plasma. Hence only tiny change in the wall retention strongly influences plasma performance and, accordingly, the density control of plasmas becomes difficult for the saturated walls. This is the reason why the low recycling or wall pumping scheme, i.e. small reemission before the establishment of the steady state, is preferred to obtain good plasma confinement. Since long pulse or steady-state operations exclude the low recycling regime, it is one of the most important tasks of ITER to confirm the steady-state operation with the saturated wall being possible.

8.4.2.2 Isotopic Replacement and Appearance of the Lightest Hydrogen Isotope (H) as an Impurity

Fuel dilution by the lightest hydrogen isotope (H) which is not the fuel cannot be avoided. Water vapor is ubiquitous and is adsorbed on any material surfaces. H atoms are included in any materials as an impurity with its concentration more than ppm order. It is also well known that water vapor always remains as a residual gas in any vacuum systems. During D fueled discharges, the reemitted D reacts with H and H_2O adsorbed on PFS and release H. Accordingly, H concentration in the D discharges decreases with discharge time. However, the release of H from PFM continues owing to isotopic replacement of H retained in near-surface regions. Since depletion of H in near-surface region induced diffusion of H dissolved in deep diffusion, emission of H from the wall continues very long.

Therefore, it has been hardly possible to make pure D discharges and some H always remains in present tokamaks. Figure 8.11 [35] shows a typical example showing how isotopic changeover proceeded when fueling gas was changed from H to D or vice versa in JET plasmas with C-wall and ILW for succeeding discharges. Although initially the isotopic changeover is first, a few % of minor isotope remains quite long. Thus in "D discharges" in present tokamak always unintentionally includes a few % of H. There is a tendency that for lower wall temperatures H concentration is higher because absorbed H_2O and dissolved H are larger.

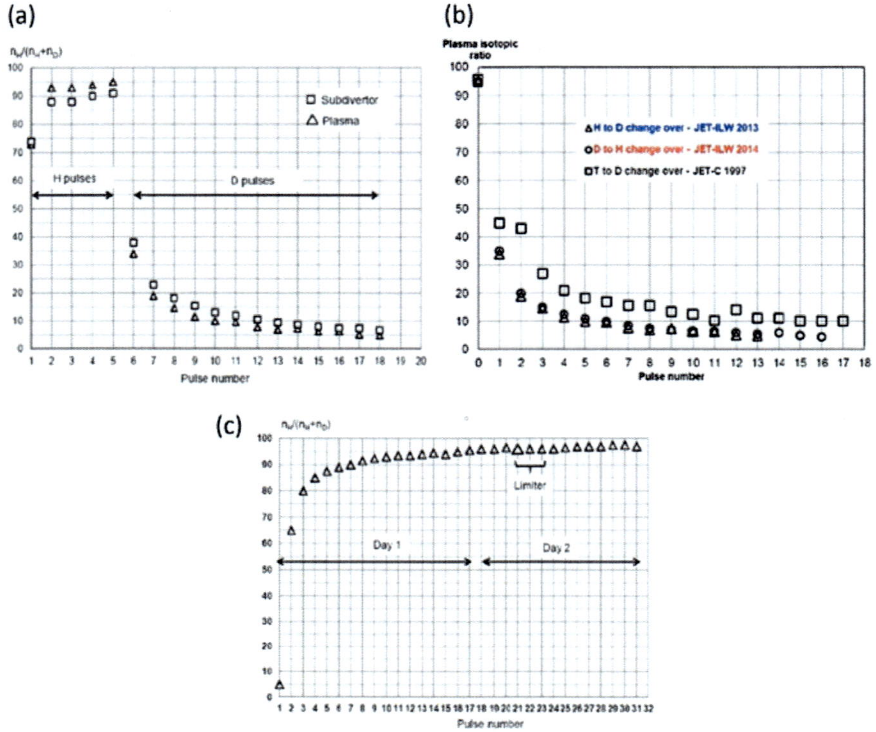

Fig. 8.11 Isotopic changeover of from H to D discharges or vice versa, **a** Isotopic ratio $n_H/(n_H + n_D)$ measured at 18 s during a low triangularity phase as a function of the pulse number, in the bulk plasma (triangles) and in the subdivertor region (squares), **b** Plasma isotopic ratio $n_D/(n_H + n_D)$ (○) for D to H changeover, $n_H/(n_H + n_D)$ (△) for the H to D changeover as a function of the pulse number for the JET-ILW and $n_T/(n_T + n_D)$ (□) for the T to D changeover for JET-CW, **c** Plasma isotopic ratio $n_H/(n_H + n_D)$ for D to H changeover as a function of the pulse number for the JET-ILW over two consecutive days of H injection (reprinted with permission from [35])

In Fig. 8.11, it is depicted that the replacement of D by H is easier than the opposite case, and the C-wall requires longer time for the isotopic replacement compared to metallic wall owing to higher hydrogen retention in C particularly its near-surface region compared to metals.

In JT-60U, after every DD discharge campaign, numbers of HH discharges were done for removal of T remained in C PFM during the DD discharge campaign. The HH discharges made isotopic replacement, removing D from near surface to deeper region and increased H concentration as shown in the depth profiles of H and D in Fig. 8.12 [36]. Characteristics of the replacement are divided into three different regimes; eroded areas with higher flux (Fig. 8.12a) (at outer divertor) and lower flux (Fig. 8.12b) (inner divertor) and deposited layers on dome in private flux region. D retained in near surface was mostly replaced by H. D penetrating deeper region owing to high temperature as seen in (a) was hardly replaced so as D in retained in the deposited layers in (Fig. 8.12c).

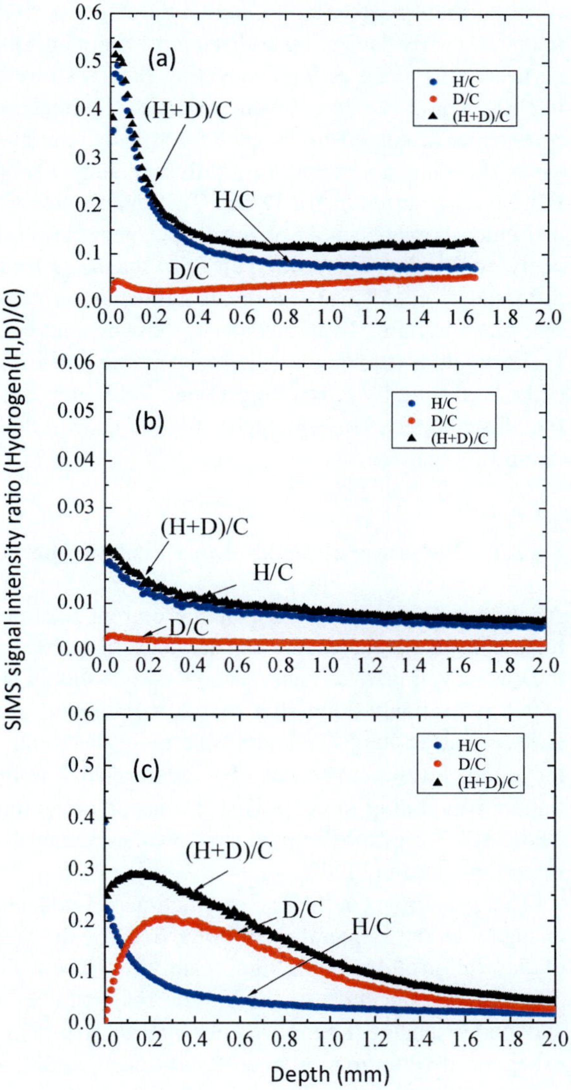

Fig. 8.12 SIMS depth profiles within 2.0 mm from the top surface for CFC tiles exposed to DD discharges followed by HH discharges in JT-60U. **a** the outer divertor, **b** the inner divertor, and **c** the dome tiles [36]

Thus, the characteristics of the isotopic replacement are quite dependent on how D is retained in depth and the temperature during the plasma exposure. It is difficult to foresee how it would appear, because higher temperature decreases H remained in PFM, while it enhances H supply from bulk. At the steady-state discharges in a reactor, one would expect significant reduction of H, while the existence of ubiquitous water would not allow this, because any surfaces of systems could be the source of H and back diffusion of H_2O in pumping system could be the source. Therefore, in ITER, the existence of a few % of H is allowed. In a reactor, further reduction seems necessary.

In burning plasma, both D and T impinge to PFM and both are simultaneously retained. At the steady-state discharges, injecting fluxes of D and T to PFS must be nearly the same as their recycling fluxes. However, D and T retention profiles in PFM would not be the same, because the injecting fluxes of D and T must not be even as discussed in Chap. 7, even if the energy of injecting D and T was the same, their injecting depths are different, and trapping energy or binding energy in PFM is also different for D and T. Probably larger diffusion or migration length of lighter element, i.e. D, gives deeper penetration than T, while smaller trapping energy or binding energy of T in PFM makes easier reemission of T than D. These differences would give significant influence on controlling D/T ratio to keep the optimized burning. Until now, no systematic studies have been done in this respect. ITER would be the first test bed to examine D and T recycling for the optimization of D-T burning by controlling D and T fueling. Most probably D and T must be fueled separately. Unfortunately, it is not easy to determine D and T concentration in burning plasma.

8.4.2.3 Recycling at Steady-State Plasma Operation

As discussed above, in most of the current tokamak experiments, low recycling regime, is preferred because density control can be simply done by fueling. To attain the low recycling, hydrogen retained in PFM should be minimized before discharges, which is routinely done after certain numbers of discharges as wall conditioning. However, after long discharge time or in the steady-state discharge, keeping low recycling is difficult because PFS is saturated with the fuel except for a small fraction of penetrating in deep. Hence, plasma operation with saturated wall or under nearly 100% recycling regime is not well established and remains as one of the most important tasks in ITER.

The recycling coefficient or fractions of reflection, reemission and retention change with the temperature of PFM. Because the saturation concentrations for both of dissolution and trapping significantly decrease with the temperature above around 500 K. However, most of the present tokamaks operated at ambient temperatures and their discharge duration are not long enough to increase the temperature of PFM except the divertor region. Accordingly, changes of the recycling coefficient were not significant. However, in JT-60U operated at higher temperature (around 400–500 K), the recycling coefficient during a single long discharge (around 30 s) became near unity; i.e. PFS was saturated during the discharge and then released flux was nearly the same as the incident or no extra fueling was required [37]. (See Fig. 6.3)

8.4.3 Long Term Fuel Retention

Because of T safety and fuel self-sufficiency in a fusion reactor, it is necessary to determine how much fuel is retained or fuel inventory in a reactor. Since integrated flux, or fluence of hydrogen into PFM in present tokamaks is much less than that expected in a fusion reactor, the estimation of the fuel inventory in the reactor should be done by simulation with a laboratory work using H or D ion beams and/or linear plasma machines or by extrapolation of the retention data in the present tokamaks. To make reliable estimation both the data in laboratory works and in tokamaks should be carefully compared. Until now, however, both do not give reliable results, mainly because of uncertainty in fluence dependence which changes with the flux and temperature of PFM.

For determination of the hydrogen retention in present tokamaks, there are two ways; one is to integrate flux differences of throughput and exhaust with discharge duration. This is very difficult to get reliable results, because the difference is very small, and the determination of the exhaust flux includes large error. The other is to make post-mortem analysis of retained hydrogen in PFM. Although the latter is basically more reliable than the former, it is not easy to analyze all PFM with larger area. In the following, details of the fuel retention are discussed.

8.4.3.1 Fuel Retention in Carbon and Metals

The reemission from carbon materials (C) includes hydrocarbons resulting from chemical sputtering and causes significant surface recession (erosion). Initially, fuel retention mostly occurs in surface and near-surface regions. Afterward, deposited layers gradually grow, and the fuel retention is dominated in the deposited C layers. In particular, the deposited C layers on plasma shadowed area continuously grow so as the fuel retention.

In case of metals, fuel easily diffuses in and out. Therefore, after stopping the injection, large amount of fuel retained during the injection is released out. Accordingly, the difference between the dynamic retention under plasma exposure and the static retention after stopping the exposure is quite large. In contrast, fuel injected in C is mostly trapped near the surface within the projected range making D-C and T-C bonds until C is saturated with D and T. Once trapped making such bonds hydrogen (D and T) is hardly released. In addition, very small hydrogen diffusivity in C inhibits its penetration into the bulk. Therefore, the difference between the dynamic retention and the static retention in C is smaller than that of the metals as appeared in Fig. 8.13. However, the porous nature of C allows hydrogen molecules to penetrate deep through open pores resulting in the depth profiles given in Fig. 9.12. This means that residual fuel gases in a reactor vessel can penetrate through the whole thickness of a carbon tile, resulting in all carbon crystallites (C grains) surfaces being saturated with the fuel. Actually, T was observed at the backside of carbon tiles used in DT discharge campaign in JET Mark-IIA divertor [38]. However, fuel particles can hardly penetrate inside of the C grains.

Unfortunately, little systematic data for D retention with metallic plasma-facing wall have been available except recent JET experiments with ITER-like wall (ILW), i.e. W divertor and Be first wall. Nevertheless, above mentioned differences between the C-wall and ILW clearly appeared as shown in Fig. 8.13 [39, 40], which shows

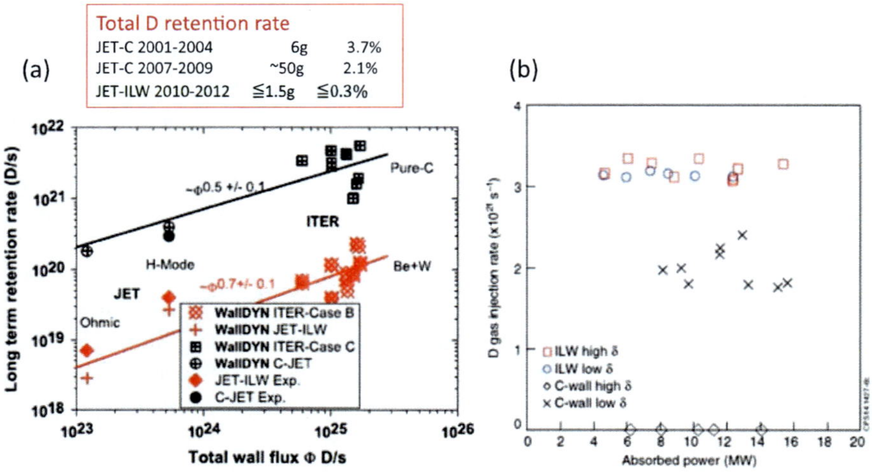

Fig. 8.13 **a** Comparison of calculated and experimental D retention rates for JET-ILW and C-JET with WallDYN calculations. Also shown are the WallDYN predictions for Be +W (Case B) and a full C-ITER (Case C) (reprinted with permission from [39]), **b** Deuterium gas injection rate during the main heating phase as a function of absorbed power for the four power scan experiments (reprinted with permission from [40])

ILW requires more hydrogen throughput than that required for the C-wall, while the static retention is large in the latter nearly 50 times. It should be noted that the retention characteristics are quite dependent on the wall temperature with less for high temperatures. In this respect, the retention in ITER would be far less than that in JET. In Chap. 9, estimation of fuel retention in a reactor is described in detail [41]. In the next section, D retention in present large machines mostly having the C-wall are summarized.

8.4.3.2 Summary of Deuterium Retention in Present Tokamaks Using DD discharges

Current estimation of T retention in ITER is based on studies of D retention in tokamaks with DD discharges (Deuterium plasma heated by Deuterium neutral beam), of which PFS is mostly composed of C. The results of these studies are summarized in this section [19–22].

In most tokamaks, as already discussed, C tiles at the outer divertor area are eroded and those at the inner divertor area deposited by the eroded carbon. However, materials balance between the erosion at the outer divertor and the deposition at the inner divertor is missing as shown in Fig. 5.1. The former is less than the latter, and the missing mass is likely compensated by erosion of the first wall of the main chamber, for which no systematic measurements have been done. The carbon deposition occurs by repetitive processes of erosion and prompt deposition of the eroded carbon as already discussed. In addition, there appears line of site deposition at remote areas from plasma; tile gaps or tile sides and divertor opening for pumping. The deposition in tile gaps is much larger for surface eroded tiles compared to that for surface deposited tiles [21], which is the evidence of prompt deposition of eroded materials.

D is incorporated in the deposited layers. But this is not due to simultaneous deposition (co-deposition) of eroded carbon with D. The concentration of D in the deposited layers is varied afterwards. Generally, injecting D flux to PFS from plasma is significantly higher than that of impurity carbon flux. Accordingly, most of injected D is spontaneously reflected and/or reemitted and only part of the incoming D remains in the deposited C layers with saturation concentration of 0.4 in D/C at maximum (see Fig. 5.7) [21]. Except far-remote area from plasma, the deposited carbon layers are subsequently exposed to boundary plasmas. And if the temperature of the redeposited carbon layers is raised, part of the retained D is desorbed. In a reactor, incident fuel flux to PFS is so large that the surface is immediately saturated with its concentration determined by the PFS temperature. In addition, the retained fuels in the saturate layers near surface are subjected to succeedingly injected fuels and part of them are always replaced. This means the fuel concentration of surface saturated layers was determined by the last exposure before the termination of discharges with little influence of previous discharges.

The situation for the eroded region is the same, i.e. the surface is saturated with D with the saturation concentration determined by the surface temperature but not the injected flux ratio of D and C. In eroded area, because of the surface recession, the thickness of the surface saturated layers would be limited only at near-surface regions and would not linearly increase with time [21]. It also suggests that the isotopic ratios of retained hydrogen near-surface layers are always equilibrated with those of incoming hydrogen isotopes fluxes (H/D/T), when different hydrogen isotopes are impinging. And the depth attaining this equilibrium is quite thick owing to the porous nature of carbon materials and is increased by temperature rise. Hence, D retention in plasma-facing surfaces (both eroded and redeposited) would be significantly reduced by isotopic replacement by HH discharges subsequently made after DD discharges in JT-60U as depicted in Fig. 8.12 [36].

In Table 8.1, D retention rates are added to the carbon deposition rates in current tokamaks [19–22]. As seen in the table, the fuel retention rates of 3–50% are found using gas balance measurements. A lower retention rate (~10%) is often obtained using post-mortem analysis of plasma-facing tiles. Since carbon erosion and deposition rates are nearly the same for most tokamaks, the fuel retention rates are totally dependent on temperatures of PFS or operation temperature owing to the temperature dependence of the saturation concentration of D in the deposited layers. Accordingly, JT-60U operated at the highest temperature (523–573 K) shows the lowest retention rate. Sometimes, JT-60U showed wall saturation, i.e. 0% retention rate.

The operating temperature of ITER, around 423 K, is still lower than the reactor operation temperature (above ~700 K) or even that of JT-60U (573 K). Accordingly, D retention in its redeposited carbon layers would be significantly large. Other tokamaks (except JET) operated at near RT show similar retention rates of 10–20%. The largest retention was observed in Tore-Supra, which had water cooled divertor/limiter blades at the bottom of its tokamak vessel [42]. In the next chapter (Chap. 9), the detailed estimation of the fuel retention in a reactor is made.

8.5 T-Related Issues on the In-Vessel T Inventory

Evaluations of the in-vessel fuel inventory have been mostly done by integration of local retentions in redeposited layers and/or near surfaces assuming toroidal symmetry in tokamaks. Although the bulk retention should have certain contribution, it has been scarcely assessed. The bulk retention in the plasma-facing materials must be very small in its concentration. Nevertheless, its integrated amount over full volume of the PFM and vacuum vessel could not be small. Here, the maximum bulk retention for metallic PFM is estimated according to crude assumptions as follows. It is well known that outgassing rates from metals are ranging from 10^{-6} to 10^{-3} $Pa{\cdot}m^3{\cdot}m^{-2}{\cdot}s^{-1}$ [43]. Taking 10^{-5} $Pa{\cdot}m^3{\cdot}m^{-2}{\cdot}s^{-1}$ as the outgassing rate from plasma-facing surfaces with their surface area of 1000 m^2, the total outgassing rate would be 10^{-2} $Pa{\cdot}m^3{\cdot}s^{-1}$. This seems negligible compared to the maximum fuel through-put in ITER of 200 $Pa{\cdot}m^3{\cdot}s^{-1}$ (~2.5 $g{\cdot}s^{-1}$) [44].

Since the source of the H outgassing is very likely H retained in the materials as impurity and H uptake can occur at the backside surface with oxidation by water. Therefore, it might be reasonable to assume tritium uptake in the bulk proceeds just opposite to the outgassing. Then about 10^{-2} $Pa{\cdot}m^3{\cdot}s^{-1}$ (0.1 $mg{\cdot}s^{-1}$) orders of tritium would be continuously retained in the bulk, i.e. every ITER discharge (400 s) would remain 0.4 g of T in the bulk and 25000 shots pile up 1 kg. This might be overestimation. However, another estimation given below would probe this T pile-up scenario is realistic considering that any materials retain H as an impurity with more than 10 ppm. Suppose T can be replaced by the impurity H, though the reaction rate is totally depending on diffusion time. Then, 10 ppm T could be dissolved in the bulk after many discharges or at steady state. This suggests that usage of 1000 tons of structure materials for a reactor could result in 10 kg of T at the maximum. This type of tritium inventory is well known in the detection of T by an ionizing chamber as a memory effect, i.e. initially introduced T to the ionization chamber and piping disappeared and certain time is required to start the detection, and the detection continues for while after stopping the T introduction as explained in Chap. 9 in ref. [33]

From the aspect of tritium safety, this type of inventory is often categorized to be immobilized T inventory and might not be a serious concern, but kg orders of T would disappear in the bulk of structure materials of all tritium handling systems at the starting phase of every reactor. This will be discussed later again.

Difficulty of quantitative analysis of T in the in-vessel components adds additional problems. It is ironical that the accuracy in detecting low levels of T (below 10^9 Bq) by its β decay is better than that in the very high levels which are determined by mass and/or pressure measurements and calorimetry with the accuracy of only 10^{-2}–10^{-4}.

In addition, one cannot detect T existing inside of solids deeper than a few μm from their surfaces. The easy isotopic exchange of T with ubiquitous hydrogen in water and hydrocarbons at materials surfaces can significantly affect the analysis and also gives concerns in safety.

8.6 Summary

Main subjects of PMI are (1) power load to PFM, (2) hydrogen retention and recycling, (3) erosion and deposition and accompanied material transport, (4) their influence on plasma confinement. Compared to a fusion reactor, injecting hydrogen flux to PFM in current tokamaks is far less. Accordingly, power load given by the fuel particles was not high enough to give appreciable influences on hydrogen recycling, erosion/deposition and their mutual relations, for example, fuel retention in deposits and formation of dusts. Hence, first three subjects have been separately studied as given in this chapter, while the fourth subject in the present tokamak is not so significant to allow systematic studies except dust formation and some changes in optical emissions in front of PFM.

In PMI in a fusion reactor, first three subjects must be strongly correlated with each other and large power load would change the appearance of PMI as suggested in the text. Furthermore, fueling T requires additional considerations on T safety, fuel self-sufficiency, and neutron irradiation effects. This chapter summarizes observations of PMI in current tokamaks with C-wall and metallic wall and suggests required studied considering significant increase of power load, incident particle flux of fuels to PFM and neutron effects.

References

1. A. Kreter, Reactor-relevant plasma-material interaction studies in linear plasma devices. Fusion Sci. Technol. **59**, 51–56 (2011)
2. J. Rapp, The challenges of plasma material interactions in nuclear fusion devices and potential solutions. Fusion Sci. Technol. **72**, 211–221 (2017)
3. R.A. Pitts, G. Arnoux, M. Beurskens et al., The impact of large ELMs on JET. J. Nucl. Mater. **390–391**, 755–759 (2009)
4. A. Huber, G. Arnoux, M.N.A. Beurskens et al., Radiation loads onto plasma-facing components of JET during transient events–experimental results and implications for ITER. J. Nucl. Mater. **415**, S821–S827 (2011)
5. N. Den Harder, S. Brezinsek, T. Pütterich, et al., ELM-resolved divertor erosion in the JET ITER-Like wall. Nucl. Fusion **56**(9), 026014 (2016)
6. G. Arnoux, B. Bazylev, M. Lehnen, et al., Heat load measurements on the JET first wall during disruptions, J. Nucl. Mater. **415**, S817–S820 (2011)
7. V. Philipps, T. Tanabe, Y. Ueda et al., Molybdenum test limiter experiments in TEXTOR. Nucl. Fusion **34**, 1417–1429 (1994)
8. R. Dux, A. Herrmann, A. Kallenbach, Plasma surface interaction with tungsten in ASDEX Upgrade. J. Nucl. Mater. **337–339**, 852–856 (2005)

9. B. Lipschultzc, Y. Lin, M.L. Reinke, Operation of Alcator C-Mod with high-Z plasma facing components and implications. Phys. Plasmas **13**, 056117 (2006)
10. Fang Ding, Guang-Nan Luo, Xiahua Chen, Plasma–tungsten interactions in experimental advanced superconducting tokamak (EAST). Tungsten **1**, 122–131 (2019)
11. V. Philipps, R. Neu, J. Rapp et al., Comparison of tokamak behavior with tungsten and low-Z plasma facing materials. Plasma Phys. Control. Fusion **42**, B293–B310 (2000)
12. G.F. Matthews et al., Overview of the ITER-like wall project. Phys. Scr. **T128**, 137–143 (2007)
13. G.F. Matthews, JET EFDA Contributors, the ASDEX-Upgrade Team, Plasma operation with an all metal first-wall: comparison of an ITER-like wall with a carbon wall in JET. J. Nucl. Mater. **438**, S2–S10 (2013)
14. I. Nunes, P.de Vries, P.J. Lomas, Optimization of the JET beryllium tile profile for power handling. Fusion. Eng. Des. **82**, 1846 (2007)
15. V. Riccardo, M. Firdaouss, E. Joffrin, Operational limits for the ITER-like wall in JET. Phys. Scr. **T38**, 014033 (2009)
16. S. Brezinsek, JET-EFDA contributors, Plasma-surface interaction in the Be/W environment: conclusions drawn from the JET-ILW for ITER. J. Nucl. Mater. **463**, 11–21 (2015)
17. K. Tani, M. Azumi, H. Kishimoto, S. Tamura, Effect of toroidal field ripple on fast ion behavior in a tokamak. J. Phys. Soc. Japan **50**, 1726–1737 (1981)
18. K. Tobita, S. Nishio, S. Konishi et al., First wall issues related with energetic particle deposition in a tokamak fusion power reactor. Fusion Eng. Design **65**, 561–568 (2003)
19. T. Tanabe, Tritium management in a fusion reactor–safety, handling and economical issues, in: *Proceedings of 2nd ITER International Summer School: Confinement*, ed. by S.-I. Itoh, S. Inagaki, M. Shindo, M. Yagi (American Institute of Physics, 2009), pp. 112–126
20. V. Philipps, Plasma–wall interaction, a key issue on the way to a steady state burning fusion device. Phys. Scr. **T123**, 24–32 (2006)
21. T. Tanabe, K. Masaki, K. Sugiyama, et al. An overview of recent erosion/deposition and hydrogen retention studies in JT-60U. Phys. Scr. **T138**(9), 014006 (2009)
22. E. Tsitrone, Key plasma wall interaction issues towards steady state operation. J. Nucl. Mater. **363–365**, 12–23 (2007)
23. A Widdowson, Material migration patterns and overview of first surface analysis of the JET ITER-like wall, Phys. Scr. **T159**(9), 014010 (2014)
24. A. Baron-Wiechec, Global erosion and deposition patterns in JET with the ITER-like wall. J. Nucl. Mater. **463**, 157–161 (2015)
25. J. Likonnen, K. Heinola, A. De Backer, Investigation of deuterium trapping and release in the JET ITER-like wall divertor using TDS and TMAP. Nucl. Mater. Energy **19**, 166–178 (2019)
26. S. Krat, M. Mayer, I. Bykovet et al., Erosion at the inner wall of JET during the discharge campaign 2013–2014. Nucl. Mater. Energy **11**, 20–24 (2017)
27. R.P. Doerner, M.J. Baldwin, D. Buchenauer et al., The role of beryllium deuteride in plasma-beryllium interactions. J. Nucl. Mater. **390–391**, 681–684 (2009)
28. J.C. Flanagan, M. Sertoli, M. Bacharis et al., Characterising dust in JET with the new ITER-like wall. Plasma Phys. Control. Fusion **57**, 014037 (2015)
29. J.P. Sharpe, D.A. Petti, H.-W. Bartels, A review of dust in fusion devices: Implications for safety and operational performance. Fusion Eng. Design **63**(64), 153–163 (2002)
30. J.P. Sharpea, P.W. Humrickhouse, C.H. Skinner et al., Characterization of dust collected from NSTX and JT-60U. J. Nucl. Mater. **337–339**, 1000–1004 (2005)
31. S. Rosanvallon, C. Grisolia, P. Andrew et al., Dust limit management strategy in tokamaks. J. Nucl. Mater. **390–391**, 57–60 (2009)
32. A. Malizia, L.A. Poggi, J-F. Ciparisse, A review of dangerous dust in fusion reactors: from its creation to its resuspension in case of LOCA and LOVA. Energies **9**, 578 (2016)
33. T. Tanabe, *Tritium: Fuel of Fusion Reactors*. (Springer, Japan, 2017), https://doi.org/10.1007/978-4-431-56460-7
34. H. Zushi, Tritium measurement II-Tritium in plasma –Chapter 8, in *Tritium: Fuel of Fusion Reactors*, by ed. T. Tanabe (Springer Japan, 2017), https://doi.org/10.1007/978-4-431-56460-7

35. T. Loarer, S. Brezinsek, V. Philipps, et al. Plasma isotopic changeover experiments in JET under carbon and ITER-like wall conditions. Nucl. Fusion **55**(8), 043021 (2015)
36. Y. Hirohata, Y. Oya, H. Yoshida et al., Depth profile and retention of hydrogen isotopes in graphite tiles used in the W-shaped divertor of JT-60U. J. Nucl. Mater. **329–333**, 785–789 (2004)
37. T. Nakano, N. Asakura, H. Takenaga et al., Impact of wall saturation on particle control in long and high-power-heated discharges in JT-60U. Nucl. Fusion **46**, 626–634 (2006)
38. T. Tanabe, N. Bekris, P, Coad, et al. Tritium retention of plasma facing components in tokamaks. J. Nucl. Mater. **313–316**, 478–490 (2003)
39. K. Schmid, K. Krieger, S.W. Lisgo et al., Quantitative modeling of fuel retention in the JET-C and JET-ILW wall configurations by WallDYN and predictions for ITER. J. Nucl. Mater. **463**, 66–72 (2015)
40. C.D. Challis, J. Garcia, M. Beurskens, et al., Improved confinement in JET high β plasmas with an ITER-like wall. Nucl. Fusion, **55**(18), 053031 (2015)
41. G.F. Matthews, Plasma operation with an all metal first-wall: comparison of an ITER-like wall with a carbon wall in JET. J. Nucl. Mater. **438**, S2–S10 (2013)
42. B.B. Pégouriéa, C. Brosset, E. Delchambre et al., Long discharge particle balance and fuel retention in Tore Supra. Phys. Scr. **T111**, 23–28 (2004)
43. D Hoffman, B Singh, JH Thomas, *Handbook of Vacuum Science and Technology* (Academic Press, 1998)
44. ITER Technical Basis 2002, ITER EDA, Documentation Series No. 24, IAEA (Vienna, 2002)

Chapter 9
Fuel Retention in a Reactor with Full C-Wall and Full W-Wall and Its Recovery

9.1 Introduction

Because of the limited resources and radio-hazardousness of Tritium (T), accounting for T fuel inventory in a fusion reactor system and its reduction/recovery are key engineering issues for establishment of a D-T fusion reactor as an energy source. Among all components and subsystems in a fusion reactor system, T retained in a reactor vessel is the largest and most difficult to account or quantify, and also hard to recover. Although T retention in other subsystems, such as exhaust, blanket, refinement, isotope separation, storage, fueling, and piping must be also quantified, it is not directly correlated to plasma material interactions (PMI) and is not discussed here. Many reviews and textbooks on T management in a fusion reactor have been published. One can refer a textbook entitled "Tritium fuel for fusion reactors" [1] as an example. Techniques for handling T have been well established. Nevertheless, the amount of T used in a reactor is quite huge compared to conventional T handling systems and no PMI data relating to T fuel are available. Hence, PMI data based on H and D are used for investigation of PMI in a reactor using D and T as the fuels. Since the mass differences among H, D, and T are quite large, their behavior in plasma material and PMI should be significantly different from each other as introduced in Sect. 7.6 in Chap. 7. Although isotope effects clearly appear in physical and chemical phenomena in plasma confinement, PMI, and behavior in materials, for examples, better confinement, lager sputtering and less permeation, for heavier isotopes, respectively. Fortunately or unfortunately, however, PMI data themselves include rather large uncertainty (or error). This makes it difficult to discuss the isotope effect appeared in PMI quantitatively. Hence some uncertainty, within a factor, cannot be excluded in the estimation of the fuel retention in reactor based on PMI data of H and D in this chapter and the isotopic effects would be considered only when the effect exceeds the square root mass dependence of $\sqrt{3}$ for H and T and $\sqrt{2}$ for D and T.

This chapter focuses on T retention in a reactor vessel in particular how the retention piles up under long operation or under steady-state operation. The estimation

T. Tanabe, *Plasma-Material Interactions in a Controlled Fusion Reactor*, Springer Series in Plasma Science and Technology, https://doi.org/10.1007/978-981-16-0328-0_9

of the T retention is based on the retention data on H and D assuming the number of each isotope is equivalent. Since the estimated fuel retention rate is so large that the recovery of retained fuel is mandatory to keep the fuel self-sufficiency and the T safety. In the aspect of T fuel self-sufficiency, easy T recovery would be a new criterion for the selection of PMI as discussed in the next Chapter.

9.2 Present Estimation of Fuel Retention in ITER

Detailed estimation of T inventory build-up in ITER has been done by Roth et al. [2–4] as shown in Fig. 9.1, using extrapolation from experimental data and modeling for the initial choice of ITER materials. Since no tokamaks have been using T routinely, T retention in a reactor should be predicted by accumulated knowledge for retentions of H and D in various tokamak experimental devices. However, the integrated flux (fluence) of H and D in the present tokamaks is far less than that of a reactor. Furthermore, data for H and D should be converted to that for D and T considering isotope effects. Additionally, as described in the previous chapter, the total fuel retention in most cases was determined from the balance between the total throughput and the total exhaust. Although the throughput can be determined precisely, the exhaust determined by the hydrogen pressure in front of a pump multiplied with its pumping speed includes large error. Hence, the estimation of the total retention as the time integration of the difference between the two would include substantial error which could be more than one order of magnitude.

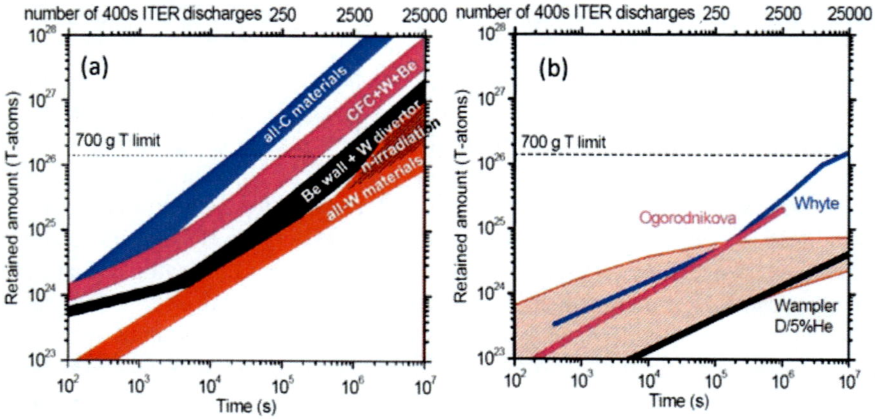

Fig. 9.1 **a** Tritium inventory in ITER for the all-C (blue line) and all-W options (red line) compared to the initial material choice (magenta). In addition, retention values for the option of a full W divertor and Be first wall are included (black line). The assessment was performed assuming different particle fluxes to different divertor and wall areas: divertor: 3 m^2, 2×10^{24} (D + T) m^{-2} s^{-1}, 775 K; 47 m^2, 2×10^{23} (D + T) m^{-2} s^{-1}, 500 K; baffle and wall: 750 m^2, 1–5 $\times 10^{20}$ (D + T) m^{-2} s^{-1}, 380–440 K. (reprinted with permission from [2]) **b** Tritium retention in ITER assessed for all-W PFCs without (shaded red area) and with neutron-damage (lines) to the wall material (reprinted with permission from [3]). See references therein

Nevertheless, these figures are used as reference until now. Figure 9.1a shows that a build-up of the tritium inventory to the administrative limit of 700 g and max 1 kg is attained within a few hundred nominal full power D/T discharges. The retention in deposited C and Be is the dominant process in C-wall and Be-wall, respectively, while bulk retention dominant in W-wall including significant increase caused by neutron irradiation (see Fig. 9.1b). The estimates have been made for the steady-state operation with many simplified assumptions. For example, the effects of ELMs and off-normal events are not considered. They will increase the surface temperature of PFS which enhances thermal release of T resulting in less T retention in bulk metals. Large uncertainties also arise, especially for a mixed material device, from complex processes such as material intermixing in deposits and/or the formation of new phases and alloys on the surface. The detailed geometrical structure of the divertor plates and vessel walls was not considered. Retention in tile gaps and at remote areas should be also added to the global inventory.

9.3 Construction of Fuel Retention Model in a Fusion Reactor

The fuel retaining areas in a reactor vessel can be divided roughly into four: (1) plasma facing area, (2) areas not directly exposed to plasma but on the line of site to plasma, (3) plasma shadowed area where nonthermalized neutral fuel particles impinge, and (4) far remote area where mostly thermalized fuel particles or remaining fuel gas impinge. First two areas are further divided into two; erosion-dominated area and deposition-dominated area. Depending on the areas, time sequences and mechanisms of T retention are different. Areas giving major T retention in most of the present tokamaks are (1) inner divertor, (2) outer dome, (3) tile gaps, and (4) deposits at remote area. In contrast, most of the first wall and outer divertor are eroded. Hence, T retention in these areas is excluded in the estimation of fuel retention in ITER given in Table 9.1 with separation of the above four areas which was made by Kirshner et al. [4] as the base for Fig. 9.1.

Table 9.1 Estimation of fuel (H + D) retention flux in an ITER scale full carbon machine operated above ~400 K. Following Kirshner et al. [4], total C deposition rate is set to be 6×10^{21} C s^{-1}

Region		C deposition fraction [%]	(H + D)/C	Fuel retention flux [x10^{20} s^{-1}]	Temperature [K]
Plasma-facing surfaces	Inner divertor	57	0.013	~0.4	1000
	Outer dome	28	0.08	~1.3	800
Shadowed areas	Tile gaps	11	0.1	~0.6	700
	Remote area	4	0.8	~2.2	423
Total		100	–	~4.6*	–

*The ratio for Plasma facing surface/Shadowed area

However, direct measurements of retained H and D in PFM and plasma-facing components (PFC) in present tokamaks have been scarcely done as already mentioned several times, except systematic studies in JT-60U with PFS of full carbon [5–10], in which direct measurements of the H and D retention in selected plasma-facing tiles located above mentioned four areas. Assuming the toroidal symmetry in the retention, the piling-up of the retention with exposed fluence was determined, which is described in the next section together with the explanation on how H and D were retained in carbon materials.

Unfortunately, JT-60U was a full carbon device in which all PFS were covered by carbon materials, and no such data are available for a full W machine. Therefore, for the full W machine, the estimation of the fuel retention was done from the balance of fueling and exhaust in some tokamak devices together with the extrapolation of laboratory experiments, which is described in Sect. 9.5.

In these estimations for the C and W and walls, the fluence of impinging fuel particles has remained still low compared to that of a reactor. Nevertheless, the comparison of estimated T retention in a C-wall machine and a W-wall machine gives important information and ideas for what shall be done to get reliable T retention in the reactor.

9.4 Fuel Retention in Carbon Materials

9.4.1 Characteristics of Hydrogen Retention in Carbon Materials [11]

Figure 9.2a shows how hydrogen (referred to as H hereafter) is penetrating and retained in carbon materials which are constructed of mainly filler (graphite like crystallite) and binder in nuclear graphite and carbon fibers (graphite-like fibers) and binder in CFC (carbon fiber enforced carbon materials) [12] and contain open and closed pores. Hydrogen molecules (H_2) can easily migrate through the open pores (Path 1), and hence C has significantly large gas permeability compared to metallic materials. H_2 also penetrates through boundaries between the crystallites (Path 2). Therefore, all surfaces of the crystallites are simultaneously exposed to gaseous H_2. H_2 can also migrate intercalantly in between the basal planes of the crystallites (Path 3), when the pressure of H_2 atmosphere is raised extremely high, which is not likely to occur in fusion environment. C atoms in the crystallites are inactive and only C atoms at their edges are very likely terminated by H making C-H bonds in hydrogen atmosphere. In carbon materials, the crystallites are not fully graphitized and include various defects which can trap hydrogen atoms. Neutron irradiation introduces additional defects in the crystallites, which also trap hydrogen to increase the retention. In normal condition (or exposure to H_2 gas), it seems very hard for H_2 to dissociate to atoms and penetrate or diffuses into the crystallites interstitially as depicted from quite small solubility and diffusivity of H in graphite.

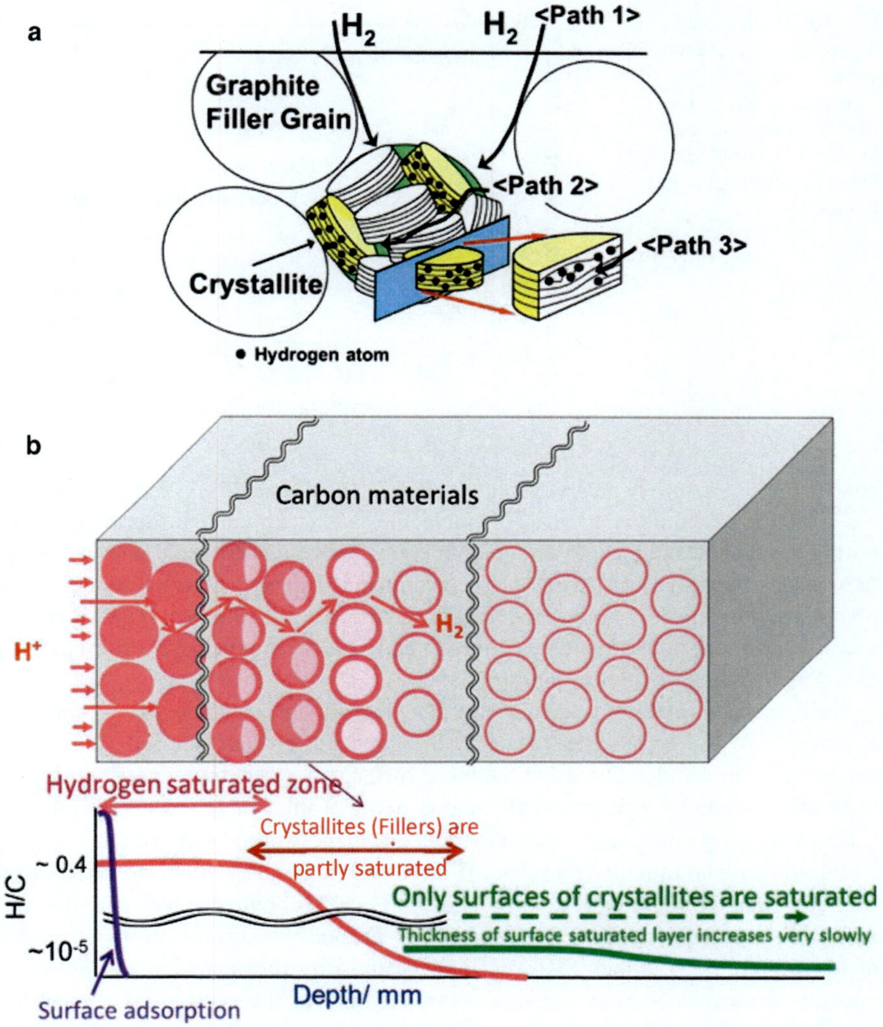

Fig. 9.2 Schematic drawings showing hydrogen retention in Carbon. **a** microstructure of carbon materials and hydrogen behavior in them [12]. **b** Schematic drawing of hydrogen uptake given by energetic H injection near RT. Circles represent fillers in C. At room temperature saturation concentration of H in C is 0.4 in H/C which decreases with increasing temperature as shown in Fig. 9.3

In case of energetic H injection, carbon atoms in the crystallites can directly react with injected H atoms and ions, or formation of interstitials, and vacancies produced by the injection enhance C-H bond formation as schematically shown in Fig. 9.2b. Accordingly, in near-surface region, hydrogen saturated layers are formed with the saturation concentration of H/C nearly 0.4 at RT and around 0.1 at 600 K as shown in Fig. 9.3 [10]. The hydrogen saturated layers having H/C larger than around 0.1

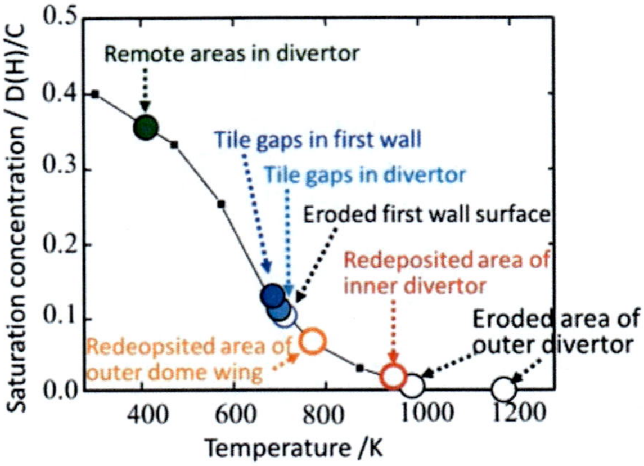

Fig. 9.3 Temperature dependences of saturated surface concentrations of H in atomic ratio (H/C) for C tiles taken from various locations of JT-60U and for H ion irradiated graphite (dotted line) [10]

change their structure from crystallites to amorphous. Some of injected H directly in the crystallites recombine to make H_2 molecules and accumulate in-between neighboring basal planes to make a lenticular opening (very thin (nm scale) but very wide (μm scale) bubble) [13]. After the surface changed to hydrogen saturated amorphous structure, the lenticular opening disappears. Instead, spherical bubbles appear because the amorphous structure is three-dimensionally homogeneous different from two-dimensional graphite crystallites.

Still some of H atoms and ions can penetrate along open pores to be retained at surfaces of the crystallites facing to the open pore, often referred to as inner surfaces. H_2 molecules formed at the inner surfaces can penetrate deeper along the open pores. Figure 9.4 schematically shows depth profiles of H and D appearing in C tiles after long time plasma exposure, DD discharges followed by HH discharges. It should be noted that different from H dissolved in metals, which can be released by diffusion after stopping H loading, most of H in C is bonded and require thermal activation (heating) to remove. The amount H + D uptake on surface of C (irrespective of loading methods, either gas exposure, plasma exposure or energetic injection) varies with degree of graphitization or crystallinity of the crystallites. Generally, poorly graphitized fillers retain more H [12] and neutron irradiation enhances the fuel retention in particular for gaseous loading [14]. Tritium profile on the eroded CFC tile of TFTR clear show different T retention between fivers and matrix (see Figs. 5.7 and 5.11 in Chap. 5). Figure 9.4 also shows how isotope replacement occurs as discussed in Sect. 8.4 in Chap. 8. D retained during DD discharges is isotopically replaced by H during HH discharges subsequently made after the DD discharges. D retained near-surface regions irrespective of the location of the surface, i.e. not only the top surface but also the surface of the crystallites or fibers facing to the open pores (the inner surfaces) are easily replaced while D retained in deeper region in the crystallites is hard to be replaced because H diffusion in the crystallites is hardly possible. At a little deeper region from the top surface, energetic T produced by DD reactions was directly injected and piled up escaping from the isotopic replacement.

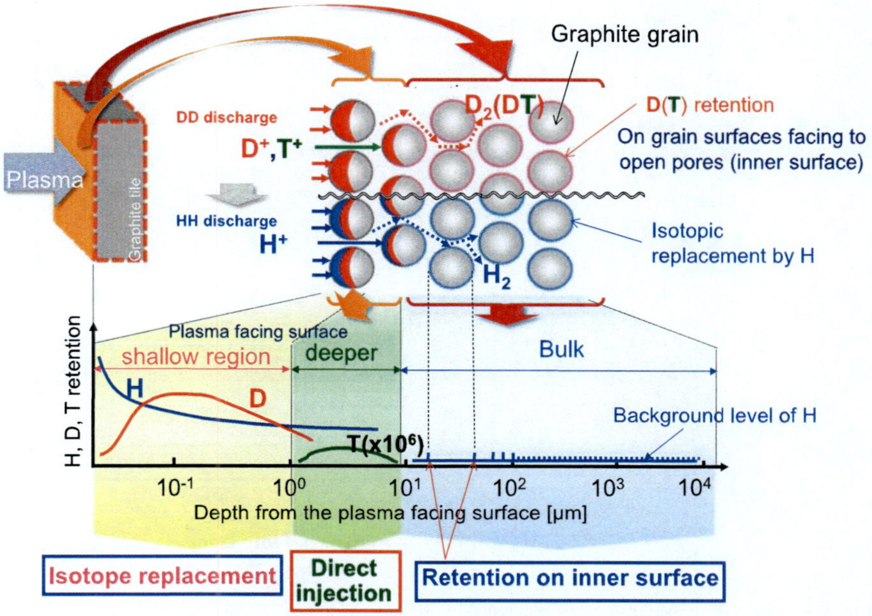

Fig. 9.4 Schematics of hydrogen uptake after exposure to DD followed by HH discharges. Circles represent fillers in C. During the HH discharges D retained in C was isotopically replaced by H from the surface. During the DD discharges, some of T produced by DD reaction is escaping from plasma and injected into a little deeper region which is not isotopically exchanged. All surfaces of fillers facing to open pores (inner surfaces) are saturated with H + D

9.4.2 Fuel Retention Build-Up in JT-60U, a Full Carbon Wall Tokamak

In JT-60U, the fuel retention has been extensively investigated in the plasma-facing carbon tiles used for divertor region and first wall, and in the deposited C layers [5–10, 15, 16]. Main discharges were done by D plasma with D neutral beam heating (referred to as DD discharges). After one experimental campaign consisting of a few hundred DD discharges, several tens of HH discharges were performed to remove T produced by DD reactions and retained in PFM. Therefore, the fuel retention includes both H and D. For the decade of the researches, data accumulation on erosion, deposition, and fuel retention in the whole vacuum vessel of JT-60U has enabled to understand the mechanisms of fuel retention and its removal method and to establish a model for the build-up of the fuels in a whole tokamak with full carbon PFM.

Figure 9.5 shows how fuel retention in JT-60U builds up with plasma exposure time [7–9]. The fuel retention is separated into 4 categories: ① deposited C layers on the plasma-facing surfaces and on the plasma shadowed areas; ② eroded areas of the first wall and of the divertor;③ retention in bulk; and in addition, ④ injection of high

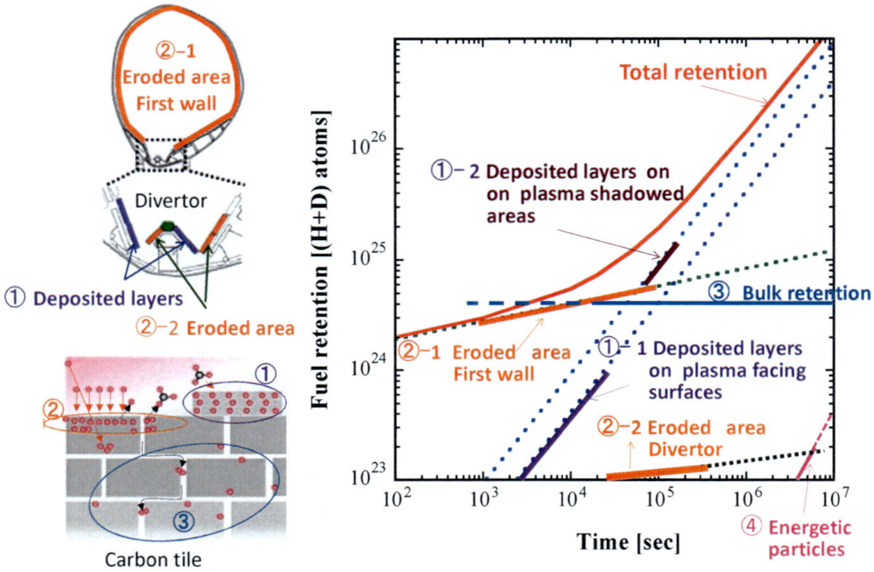

Fig. 9.5 Increase of fuel retention in JT-60U with plasma exposure time with separation into ① deposited layers on plasma-facing surfaces and plasma shadowed areas, ② deposited layers on plasma-facing surface and plasma shadowed areas, ③ bulk-retention, and ④ directly injected high energy particles

energy triton and deuteron remain a little deeper region contributing to the retention a little. Retention at far remote area is quite small in JT-60U. Before plasma exposure, carbon tiles contain some H on their surface and inner surface as described in the previous section. Although this bulk retention initially dominated with H subjected to isotopic replacement with D, its total amount of H + D is hardly increased with plasma exposure time. In early times of the plasma exposure, the fuel retention in the eroded surface dominates the total retention. Because plasma flux to the eroded area (most of the first wall) is higher and the area is wider than those of the deposited area. With increasing the plasma exposure time, the fuel retentions in the deposited layers on the plasma-facing surfaces and the plasma shadowed areas become dominant. The retention in the deposited layers on the plasma shadowed area exceeds that on the plasma facing area, because the temperature of the deposited layers on the plasma shadowed area is much lower resulting in higher saturation concentration than those of the deposited layers on the plasma facing area which is also subjected to some erosion by the plasma exposure. Furthermore, larger total area of the plasma shadowed areas (~200 m^2) than that of the plasma facing surface in the divertor regions (~7 m^2) contributes to the higher fuel retention rates. Although direct injection of energetic particles (mainly T) escaping from plasma is not large, it continues to pile up linearly with the plasm exposure time and hardly replaced by isotopic exchange because of deeper injection as shown in Fig. 9.4.

Table 9.2 summarizes carbon deposition rates and fuel retention rates in JT-60U [9, 10]. In the table, the temperature of the deposited layers during the plasma discharges are indicated. JT-60U was usually operated at around 523–573 K, the temperature of PFS, in particular divertor area appreciably increased. In Fig. 9.6, carbon deposition and fuel retention in each area are compared.

The integrated carbon deposition rates over all plasma shadowed areas (6×10^{20} C s^{-1}) were smaller than that for the carbon deposited layers on the PFS in divertor tiles, 1×10^{21} C s^{-1}. On the other hand, the fuel concentration in the former is much larger than that in the latter, because of lower temperature. The fraction of the C deposition rate on the plasma shadowed areas to the total amount of carbon deposition rate was 37%, and was smaller than that in the other devices (~50%)

Table 9.2 Summary of carbon deposition rates and retention rates of fuels in the deposition at various locations in JT-60U [9, 10]

Location	Plasma shadowed areas or Plasma facing surfaces	Temp [K]	(H+D)/C [−]	C deposition rate [Cs^{-1}]	Fraction of C deposition [%]	Fuel retention rate [$(H+D)s^{-1}$]	Fraction of fuel (H+D) retention [%]
First wall	Plasma shadowed areas (tile gaps)	700	0.1	4.0×10^{20}	25	4.0×10^{19}	30
	Plasma shadowed areas (tile gaps)	700	0.1	1.5×10^{20}	9	1.5×10^{19}	11
	Plasma shadowed areas (remote areas)	423	0.8	0.5×10^{20}	3	4.3×10^{19}	32
Divertor	Plasma facing surfaces (inner divertor tiles [10])	1000	0.013	6.7×10^{20}	42	0.8×10^{19}	7
	Plasma facing surfaces (outer dome tiles [10])	800	0.08	3.3×10^{20}	21	2.6×10^{19}	20
	Total			16.2×10^{20}	100	13.4×10^{19}	100

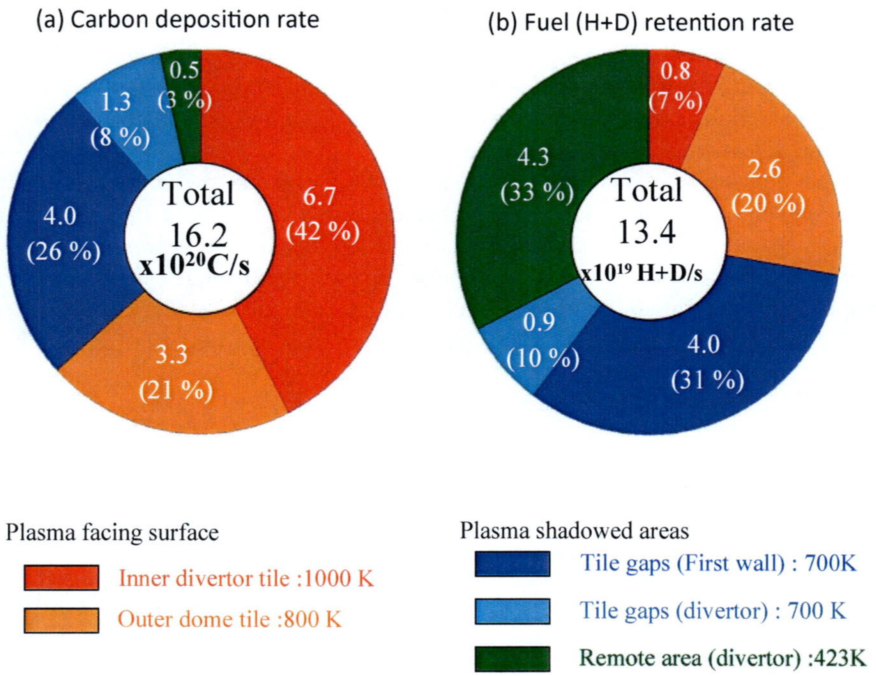

Fig. 9.6 Comparison of C deposition rates and fuel retention rates with separation of plasma-facing surface (inner divertor and outer divertor), plasma shadowed areas; tile gaps both in first wall and divertor, and remote area (bottom of divertor region) observed in JT-60U

because of divertor configuration and/or temperature in the divertor structure [17–19]. It should be noted that deposition on the first wall was not accounted because most of PFS of the first wall was eroded.

Total fuel retention rate (9.8×10^{19} (H + D) s^{-1}) of the shadowed areas by summing the fuel retention rates in tile gaps of the first wall tiles, in the tile gaps of the divertor tiles and in the remote areas underneath the divertor structure was two times higher than that (3.4×10^{19} (H + D) s^{-1}) in the plasma-facing surfaces of the divertor tiles).

Thus, the total fuel retention rate in JT-60U after long-term plasma exposure was dominated with deposited layers on the plasma shadowed area including tile gaps reaching 1.34×10^{20} (H + D) s^{-1} and fuel to carbon atom ratio is 0.08 ($= 1.34 \times 10^{20} \div 16.2 \times 10^{20}$) in (H + D)/C. This fuel retention rate was much lower than those estimated by the simulation for a full C reactor given in Fig. 9.1, mainly because of the higher temperature operation in JT-60U than the other devices used for the estimation.

9.4.3 *Estimation of Carbon Deposition and Fuel Retention in an ITER Scale Full Carbon Reactor Operated at Around 600 K*

Based on the accumulated data on carbon erosion/deposition and retention of H + D in PFS, tile gaps, bulk and deposited carbon layers in JT-60U, the fuel retention rates in an ITER scale tokamak with full carbon PFM (referred to as full-C) operated at around 600 K was estimated in the following.

The temperature rises for the full-C reactor during the plasma discharge were assumed to be the same as observed in JT-60U indicated in Table 9.2. Both the deposited C and eroded PFS were assumed to be saturated with fuel and the saturation concentration is given by Fig. 9.3. Since plasma heat load in ITER is much larger than that in JT-60U, the estimation based on the fuel retention rates determined for JT-60U may lead to overestimation of the fuel retention rates of the ITER scale full-C reactor.

The observed fluence dependence of the fuel retention in each area is extrapolated to expected higher fluence in ITER as follows.

9.4.3.1 Retention in Eroded Area

Particle fluxes at the divertor region and the first wall including baffle plates were calculated to be $1 \times 10^{24}\ \mathrm{m^{-2}\,s^{-1}}$ and $1 \times 10^{20}\ \mathrm{m^{-2}\,s^{-1}}$, respectively, and the retention rate at the eroded surface was assumed to be the same as that for the JT-60U. The fluence dependences of the fuel retentions of the outer divertor region and the first wall were assumed to be $\Gamma^{0.13}$ and $\Gamma^{0.16}$ as observed in JT-60U, respectively [10]. The surface areas of the divertor area and the first wall are roughly 55 $\mathrm{m^2}$ (25 $\mathrm{m^2}$ for the outer divertor and 30 $\mathrm{m^2}$ for the dome) and 750 $\mathrm{m^2}$ including 50 $\mathrm{m^2}$ for the baffle plate, respectively. Consequently, the estimated retention rate in the whole eroded area is around $2 \times 10^{20}\ \mathrm{s^{-1}}$ as given in Fig. 9.5, which largely contributes to the total fuel retention only in early discharge time.

9.4.3.2 Retention in Deposited Area

Using the estimated C erosion rate for full-C ITER, 6×10^{21} C $\mathrm{s^{-1}}$ [4], the carbon deposition rates in plasma-facing surfaces and plasma shadowed areas were calculated by multiplying the fractions of carbon deposition in each region observed in JT-60U. Hydrogen concentrations in deposited layers were determined from Fig. 9.3, to be 0.013 for the inner divertor tile above 1000 K, 0.08 for the dome tile at ~800 K, 0.1 for the tile gaps in the first wall above 700 K, 0.1 for the tile gaps of the divertor tiles at above 700 K and 0.8 for the remote areas of the divertor structure at 423 K. By integrating carbon deposition rates multiplied by the hydrogen concentrations in each region, the total fuel retention build-ups were estimated to be ~$5 \times 10^{20}\ \mathrm{s^{-1}}$ for ITER scale full-C reactor.

9.4.3.3 Bulk Retention

For carbon tiles, bulk retention does not likely change with fluence significantly. Because the bulk retention is mostly caused by surface and near-surface region in crystallites or filler particles (grains). Once tiles were exposed gaseous hydrogen, most of their surface is saturated and penetration inside by diffusion is hardly possible. Therefore, the contribution of the bulk retention dominates only earlier discharges. The hydrogen concentration in the bulk of carbon tiles at the eroded divertor regions and the first wall are around to be 2×10^{-4} and 1×10^{-5}, respectively, according to the data obtained in JT-60U.

9.4.3.4 Direct Injection of Energetic Fuel Particles

Based on the result that 13% of T produced by the DD reactions was retained in the outer first wall, the annual amount of T retention in the whole outer first wall was estimated. On the assumption that 17.6 MeV of fusion energy produced by one DT reaction are fully converted to useful power supply, about 20 kg of T and 13 kg of D are required to be DT reaction annually in ITER. In ITER, the DD reactions simultaneously occur with the DT reactions. At the plasma temperature of about 10^9 K, cross section of the DD reaction is around 10^{-1} of that of the DT reaction. This means that about 1.3 kg of D will be used for the DD reactions to produce about 1 kg of energetic T. Applying the value of 13%, about 130 g of T at maximum will be retained annually in the whole outer first wall. Stronger magnetic field and less ripple loss of ITER would significantly reduce the value. Still the T retention by the direct implantation would be a few tens gram. Since the direct implantation piles up and is hardly removed, this type of T retention contributes after several years of operation.

9.4.3.5 Total Fuel Retention

Adding all the above-estimated fuel retention, the total fuel retention in ITER scale full-C reactor operated above 573 K becomes as shown in Fig. 9.7b. In the figure, the retention is normalized with 100 m^2 for easy comparison. For a 3MWth reactor, the total retention would be 3 times larger. After long time operation, the fuel retention in the deposits dominate, and the total fuel retention build-ups were estimated to be ~5×10^{20} s^{-1} which is mostly owing to the retention in deposited layers on plasma shadowed area. As seen in the figure, the total inventory for full carbon stays below the safety limit of 700 g T for more than 1000 discharges and are smaller by a factor of ~10 than those for the full C-wall ITER's estimated by Roth et al. [2] in Fig. 9.1a. This is simply because the temperature of PFM in the reactor is around 150 K higher than ITER. In Roth's estimation, the saturation concentration of $(D + T)/C = 0.4$ in C deposited layers was employed, because of lower temperature operation of ITER. But in the estimation based on JT-60U, the saturation concentration in deposited C

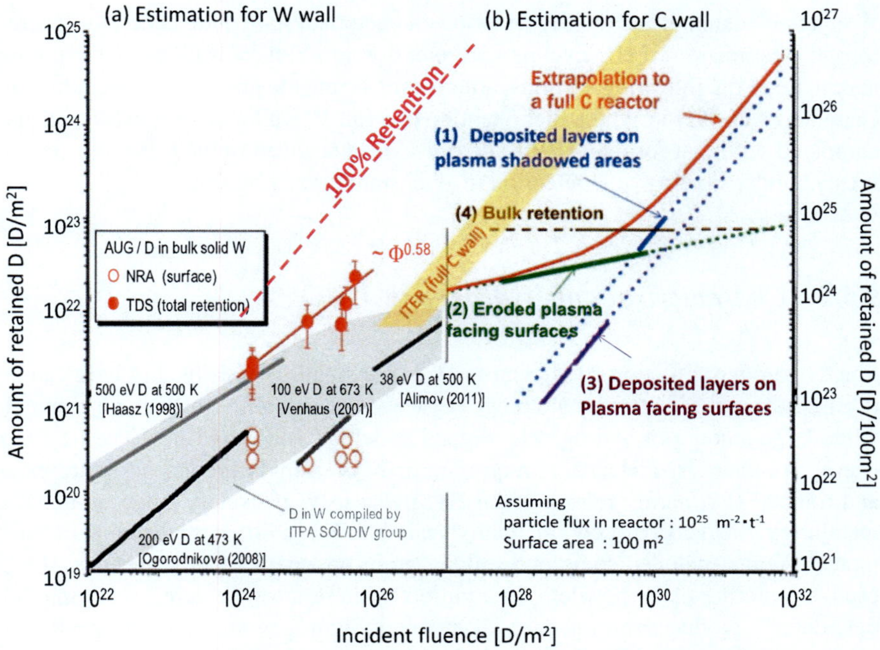

Fig. 9.7 Comparison of fluence dependence of fuel retention for W-wall and C-wall. **a** Deuterium retention in W exposed D plasma (ADEX-U) and D ions in various laboratories (reprinted with permission from [26]) and references therein, and **b** Estimation of fuel retention in an ITER scale full-C reactor [10]. The vertical scales of left and right are given by 1 m^{-2} and 10^{-2} m^{-2}, respectively, for easy comparison

layers was taken to be T/C = 0.013 and 0.08 for the inner divertor and the outer dome wing, respectively, instead of 0.4. T/C for the tile gaps was assumed to be 0.1–0.08. Since a higher tile temperature above 600 K is foreseen in ITER and future plasma devices, the fuel retention could be lower than the present estimation for ITER.

Based on the estimation given above, T retention could be significantly reduced even using C as PFM, if tile temperature of a reactor was raised above 500 K as JT-60U, it would significantly reduce the fuel retention even using full-C as PFM.

9.5 Fuel Retention in Tungsten (W)

Because no tokamaks with full W-wall are available, estimation of fuel retention in a full W-wall reactor is quite hard and extrapolation of existing data on hydrogen retention under energetic ion injection or plasma exposure to reactor like condition is necessary. However, data of hydrogen retention in W are widely scattered and the retention mechanism is not fully understood. Furthermore, formation of deposited

W layers in tokamak and their H retention characteristics are quite uncertain. Therefore, the estimation of H recycling and retention in a full-W wall reactor are quite uncertain. In the following sections, after summarizing the present understanding on characteristics of H in W, the fuel retention in a full W-wall reactor is estimated and compared with that for the ITER scale full-C reactor given in the previous section. Present understanding of H retention in W is summarized by Tanabe [20].

9.5.1 Characteristics of Hydrogen in W

Since hydrogen diffusion coefficient in W is not small, H dissolved in bulk can be released even at RT after long storage. However, H solubility in W is quite small, while large amount of hydrogen is trapped at defects and impurities. The trapping energy is rather large. Hence, H migration in W is strongly influenced by trapping and trapped H is hardly released near RT. Because W is usually made by powder metallurgy, the density of commercially available W is a little less than that of fully melted W and included some pores along grain boundaries and impurities like C used at sintering of W powders. In addition to these intrinsic defects and impurities, defects produced by injection of energetic hydrogen and neutron irradiation work as trapping sites. The fuel trapped at defects produced by neutron irradiation and distributed throughout W bulk contribute as T inventory. In addition, defects produced by energetic hydrogen injection accumulate near surface and significantly increase H retention of PFM making H saturated surface layers with H/W of ~0.1. The thickness of the H saturated layers grows with H fluence. Accordingly, H retention in W used as PFM is not likely much smaller than that in C owing to the bulk trapping and the formation of H saturated layers in near suface region.

Figure 9.8 schematically shows how H injected into W is retained. Different from C, W is strongly influenced by power load. In particular, PFS subjected to high-power load is recrystallized or their grain sizes grow, as indicated in the figure. In addition, H retention with very high concentration completely changes the crystalline structure. Near room temperature hydrogen blisters are formed and their exfoliation could result in heavy surface erosion. Depth profile of H retained in W under exposure of H plasma are schematically described in Fig. 9.9 [20]. More than 1 mono-layer of adsorbed H (around $10^{19}/m^2$) always exists on the top surface. Beneath the top surface, H saturated layers are formed. Although the amount of retained H in the H saturated layers with the concentration of around 0.1 in H/W ratio changes with the incident energy of H, most of laboratory experiments show saturation in near-surface region with $5 \times 10^{23}\ m^{-2}$ (see Fig. 9.11). Since the thickness of the saturated layers can grow with the plasma exposure time, H retention near-surface region would further increase. In addition, H can be easily diffuse into deep showing exponentially decaying concentration profile from the surface. Since the H solubility is very small, the contribution of the dissolved H in the deeper region would not exceed trapped H in near-surface region. However, defects produced by neutron irradiation could have large contribution [21]. Fluence dependence of H retention in W is given again in the next section.

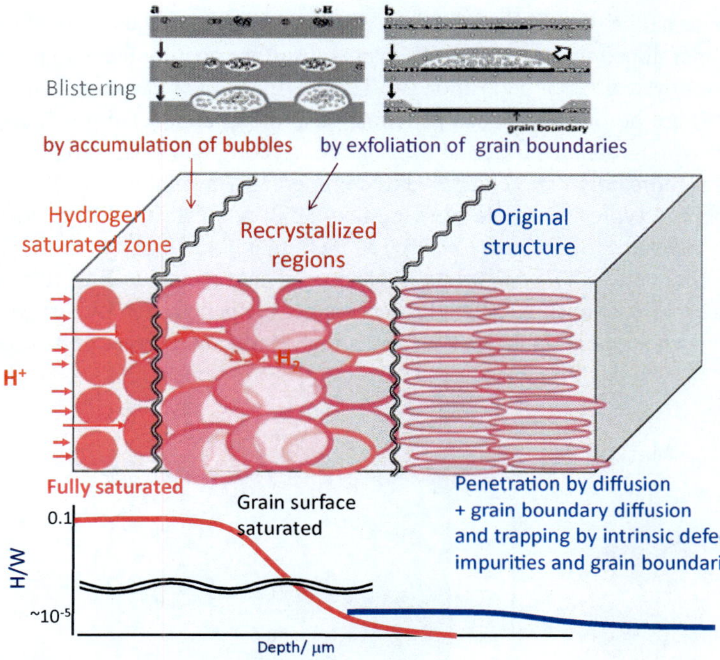

Fig. 9.8 Schematic drawing for retention of H injected in W. Top surface and subsurface are saturated with H. A few to several μm from the top surface should be recrystallized owing to high-power load, and surface and subsurface regions should be saturated with H. Hydrogen diffuses into bulk to make exponentially decaying concentration profile and without appreciable grain boundary diffusion

Fig. 9.9 Schematic depth profile of H injected into W with estimated areal or volume densities [20] based on Fig. 9.8

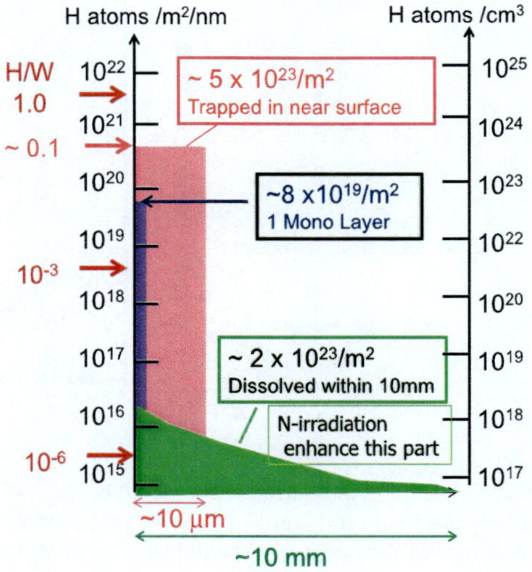

As an alternative of bulk W, deposited W on heat sink materials is considered. It should be mentioned that deposited W layers by surface coating techniques such as chemical or physical vapor deposition (CVD, PVD), Argon or vacuum plasma spray (APS, VPS) are porous and H can penetrate deep along connected pours and gran boundaries to saturate all surface of every grain. Accordingly observed H retention in them is significantly large compared to commercially available bulk W [22, 23]. Figure 9.10 is a typical example showing how H(T) retained in VPS-W on F-82H steel after exposure to H plasma or H_2 gas including T [24]. For both cases, T is uniformly retained in VPS-W, while T penetrated into substrate F-82H by diffusion. Furthermore, exposure to the plasma enhances H penetration by diffusion and resulted in even permeation in F-82H, suggesting significant increase of T inventory.

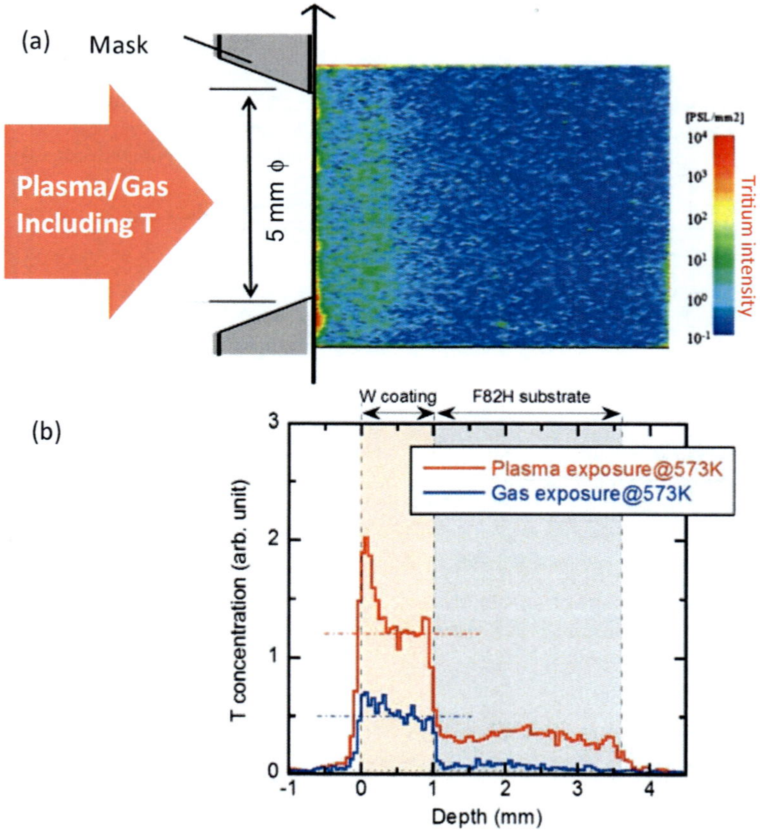

Fig. 9.10 Tritium distribution in depth of VPS-W-coated F-82H steel after exposure to H plasma or H_2 gas including T at 573 K. **a** Tritium profile after plasma exposure and **b** depth profiles after plasma and gas exposure. The thicknesses of VPS-W and F-82H are 1 mm and 3 mm, respectively [24]

9.5.2 *Fluence Dependence of H Retention in W*

Observed H retention in W given by ion injection or plasma exposure at temperature range of 380–643 K are summarized in Fig. 9.11a [25]. Large data scattering more than two orders of magnitude are appearing. The reason for such large scattering is not clear, most probably owing difference in trapping depending on W materials used, methods of H exposure, or energy of injected H.

The results obtained in ASDEX-U is given in Fig. 9.11b [26] for comparison, which is the same figure used in Fig. 9.7a. It should be noted that significant difference appeared in the retention determined by thermal desorption spectroscopy (TDS) and nuclear reaction analysis (NRA), because the latter detects H only in near surface while the former detects all. For long-term H retention in W, H dissolved and trapped in deeper region dominate the whole retention. If the retention was controlled by diffusion, the square root flux dependence should appear. The observed flux dependence is the power of 0.58, near to 0.5, supporting diffusive dissolution dominating the long-term retention. This suggests less inhomogeneity in retention profile of W-wall on the location because the contribution of H retention in deposits should be small. Thus, it is possible to compare the H retention normalized with unit area of W with that in the full-C wall in Fig. 9.7.

Since the maximum fluence in Fig. 9.11 is still far less than that expected in a reactor, the experimental data should be extrapolated to higher fluence. The fluence dependences for all data are within $\phi^{\sim 0.3}$ and $\phi^{\sim 1.0}$. Moreover, these data obtained

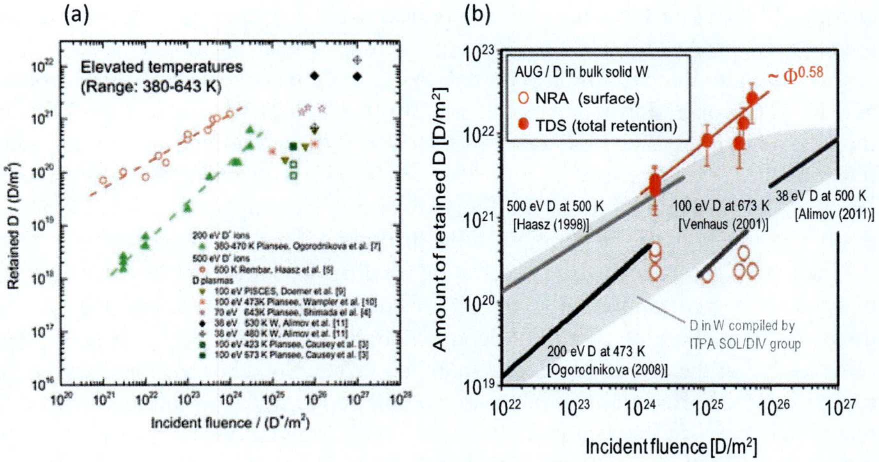

Fig. 9.11 **a** Observed H retention in W given by ion injection or plasma exposure at temperature range of 380-643 K (reprinted with permission from [25]) and references therein and **b** The results obtained for W samples exposed to ASDEX-U plasma (reprinted with permission from [26]). The results determined by thermal desorption spectroscopy (TDS) and nuclear reaction analysis (NRA) are compared. Figure **b** is used in Fig. 9.7 for comparison with the H retention in C

by laboratory experiments are quite consistent with recent data obtained for plasma-facing W-coated C tiles used in ASDEX-U [26]. One should note that the highest total retention observed is about 2×10^{22} m^{-2}, which is still far less than the ITER safety limit of 1.5×10^{24} m^{-2} (assuming 750 g is distributed homogeneously on 100 m^2 plasma-facing surface). Data extrapolation to higher fluences is required to estimate T retention in ITER and DEMO. It is critically important to know whether the retention saturates or not. Although simple extrapolation following the fluence dependences of $\phi^{0.3\sim1.0}$ would not give saturation, some data suggests saturation [28].

To make the correct estimation or extrapolation, we have to know depth profiles in deeper regions, i.e. the micrometer range is not enough but more than cm depth is required. As demonstrated in modeled depth profiles H implanted in W shown in Fig. 9.8, the depth profile can be separated into three components: (1) the trapping at the topmost surface, (2) formation of saturated layers in subsurface, and (3) bulk retention in depth. The trapping at the topmost surface always exists but would not exceed a value corresponding to a few times of surface density of W atoms, i.e. ~10^{20} m^{-2}. Significant amount of H is accumulated or trapped to make the H saturated layers in the subsurface. Hydrogen concentration in the saturated layers would not increase but their thickness could continuously increase with fluence, though the fluence dependence of the growth rate of their thickness is not clear. The saturation concentration decreases with temperature, while it increases with the H fugacity, or with flux and incident energy of loaded H. Since both temperature and the fugacity must have gradients from the surface to the bulk, the concentration in the saturated layers should not be homogeneous. Thus, to describe retention phenomena correctly, the concept of fixed trapping energies would not be suitable but multiple trapping energies [22, 29], or thermodynamic approach using free energy, or entropy effect are required.

According to the data so far reported, the saturation concentration in W around 400 K, ITER operation temperature, is 0.01–0.001 in H/W atomic ratio [25]. If the thickness of the saturated layers with H/W = 0.01 was around 10 μm, the total retention in the saturated layers would be ~6×10^{21} m^{-2}, still two orders of magnitude less than the ITER safety limit. But longer exposure with higher flux in ITER could easily increase both the H/W ratio and the thickness of the saturated layers.

The retention caused by deep penetration via diffusion is also not small. Assuming that the surface concentration of the dissolved H is around 100 appm, the total H dissolved in the bulk of 10 mm in depth would be ~10^{22} m^{-2}, which is nearly the same as that in the above-mentioned saturated layers. Somehow, H dissolved in W can be released after the exposure and isotopic replacement would also reduce T retention in near-surface region. However, the deep penetration always accompanies trapping and only tiny amount of H dissolved in W could be released. Hydrogen isotopes can reach to the back surface (permeation) which is a concern for tritium safety.

How does such H uptake modify the structure of near-surface layer of W? The answer is given in Sect. 6.3.3.2 in Chap. 6. Nevertheless, it should be noted here again, referring to Fig. 6.11 [27]. Here, H trapped or adsorbed at the topmost surface was not considered. To form the H saturated subsurface layers, firstly all grains at the geometrical surface are saturated with H and then the surfaces of grains beneath the saturated grains are saturated. Subsequently, the thickness of the saturated layers of the grains beneath the saturated grains grows until whole volume of these grains is saturated. The increase in the number of saturated grains results in the growth of the thickness of the saturated layers as shown in Fig. 9.8. Higher H flux loading would increase the H concentration in the saturated layers. Simultaneously, surface temperature rise due to the H loading reduces the saturation concentration. In addition, the temperature rise would enhance grain growth and H penetration as well, and consequently accelerates the growth of the thickness of the H saturated layers. In SIMS depth profiles [30], hydrogen retention in form of H_2 molecules are reported, which is quite natural as the cause of blisters and could be one of the causes of large data scattering in the H retention data in Fig. 9.11.

It is noted that the H saturation modifies mechanical properties and oversaturated H would precipitate as H bubbles. As a result, blistering with two different mechanisms appears (1) accumulation of gas bubbles occurring in very near-surface layers with smaller radius and high pressure inside and (2) exfoliation of grain boundaries at the deeper region of the saturated layers with/without slipping of particular plane (most probably [110] plane owing to the compressive stress in the H saturated layers) as shown in Figs. 9.8 and 9.9. The formation of the H saturated layers might be different from simple trapping but can be viewed as the formation of a chemical compound and the H release is something like the decomposition of the compound. Still the non-saturation behavior in long-term retention cannot be interpreted simply by the formation of near-surface saturated layers and their growth. Deep penetration by diffusion and immobilization by trapping in bulk must give significant contribution. Neutron irradiation creates trapping sites in deeper region, which trap H penetrated deep by diffusion to be immobilized and contribute to the non-saturation. This could result in significant H retention in W which is hard to remove and dissipate the advantage of W as PFW.

9.6 Comparison of Estimated Fuel Retention in a Reactor with Full C-Wall and W-Wall

In Fig. 9.7, fluence dependences of the fuel retention rate of the C wall and the W wall are compared, assuming surface area of 100 m^2 and particle flux of 10^{25} $m^{-2} \cdot t^{-1}$ for the former, while for latter the retention data normalized in unit area (m^{-2}). The comparison would not be straightforward because the estimation for the W wall is just the extrapolation of the data obtained for W plates in laboratory or probe samples set in ASDEX. Nevertheless, the comparison in Fig. 9.7 seems reasonable because the H retention in W is dominated in bulk and little influence on erosion and deposition. It should be noted both overlap rather well. And the fuel retention rate could be kept 1% or less for reactor scale fluence. However, owing to large error caused by the extrapolation, it is hard to foresee the fuel retention rate of 0.1% or below to attain the fuel self-sufficiency.

However, the fuel retention rate is not necessarily constant with fluence and could change from nearly 100% (except direct reflection) to 0% and the fuel retention rate should be quite small if PFM was saturated with the fuel, i.e. no more retention in the saturated wall. The initial fuel retention rate is very large because PFM is empty and most of injected fuels are retained, while for saturated PFM with fuels most of injected fuels are reemitted. Because of higher temperature operation of a reactor, the saturation concentration is much less than that observed in most of present tokamaks operated at ambient temperature without cooling. Consequently, the surface saturation will be attained at less fluence. Actually, the surface saturation appeared in JT-60U resulting in loss of wall pumping as shown in Table 8.1.

Thus, the fuel retention rate changes with fluence as appeared in its fluence dependence. Only when it shows linear flux dependence, the fuel retention rate stays constant, otherwise it changes with fluence. The linear fluence dependence appears for the fuel retention in deposited layers at plasma shadowed area for the C-wall, because of no erosion. When the fuel retention is caused by penetration (diffusion) into inner side, the flux dependence deviates from the liner one, depending on penetration mechanism it would change.

In the fluence dependence of the fuel retention given in Fig. 9.7, C and W clearly show difference. For W, the retention in deposited layers is not included, while the retention in carbon wall is dominated in the deposited layers. Accordingly, the fluence dependence for W is between $\phi^{0.5}$ and $\phi^{1.0}$ and that for the carbon wall nearly $\phi^{1.0}$ for higher fluence. Although the fluence for W stays still lower level compared to the carbon wall, the effect of its surface saturation appears as downward deviation from ϕ^1 at higher fluence. In a reactor with the full W-wall, the fuel retention in the deposited layers would be less than that in the bulk. Therefore, to keep the fuel retention rate to be less than 0.1% seems possible in both C-wall and W wall. However, it should be mentioned that neutron irradiation could give the upward deviation due to H trapping at defects produced by the irradiation.

Thus, the fuel retention rate could be reduced to be less than 0.1% at the steady-state operation. However, integrated fuel retention continuously increases. Owing to T regulation low, the maximum allowable T inventory in a rector will be set (in ITER it is 1 kg), which requires occasional T removal from PFM in addition to necessary T recovery to attain fuel self-sufficiency.

9.7 Fuel Removal/Recovery

Recovery of T retained in PFM is directly connected to T safety and T regulation. Furthermore, for establishment of fuel self-sufficiency, the recovery should be as much as possible. Although variety of techniques have been developed for T removal, they are mainly for detritiation of contaminated surfaces [31]. However, considering reuse or recycling of T in a fusion reactor, techniques would be limited, i.e. recovered T should not be diluted by H or other gases.

The most reliable method is thermal desorption at higher temperatures. However, all components in a reactor are difficult to take out to outside and require on-site heating (main pumping system with T processing systems are always available). The temperature required for full recovery is quite high and make all components to be such high temperature is not realistic. Probably, baking the whole tokamak system for conditioning to get good plasma assists T removal, but it would not be enough. Hence, various techniques are proposed and studied [32].

Methods are divided into three categories, simple out gassing or heating, mechanical removal together with materials retaining T including oxidation, and isotopic exchange.

For example, careful attention to optimizing the tritium fueling efficiency (minimizing the tritium introduced), tailoring the isotope ratio [33], and ending every shot in a deuterium-only phase with sweeping of the divertor strike points [34] have all been proposed. Seeding the divertor plasma with nitrogen may also help reduce retention rates, either by increasing the chemical erosion of tritium-containing hydrocarbon co-deposits [35] or by reducing the availability of hydrocarbon radicals, which suppresses deposit formation [36, 37]. This so-called nitrogen scavenging effect may even reduce the formation of the deposits in shadowed regions of the divertor far from the plasma [38], although issues such as nitrogen leakage out of the divertor and possible collateral damage effects have yet to be fully assessed for ITER conditions. Wall conditioning techniques, such as low-pressure RF discharges in deuterium, could further contribute to reducing the rate of tritium accumulation. Although the efficiency of this type of technique would appear not to be very high [34], it has no deleterious impact on the vessel conditions and can therefore be employed frequently (i.e. inter-shot). During the maintenance phase, it may therefore be necessary to employ other, more invasive recovery approaches. Surface heating using scanning lasers [39, 40] or flash-lamps [41] under remote handling may be an important option during this phase, for example, to remove tritium accumulated near surface of CFC which needs higher temperatures than are easily accessible with baking, up to ~750 K [42]. Since characteristics of fuel retention is different between C and W, fuel removal or recovery is introduced in the following separately.

9.7.1 Removal/Recovery of T Retained in Carbon Materials

Since very high temperature (energy) is required to desorb or remove T retained in carbon materials, removal or recovery of T retained in plasma-facing carbon tiles is concerned both for T safety and fuel recovery to attain fuel self-sufficiency. Although various methods have been examined in laboratory and tokamaks, no methods have been established as efficient or reliable [43]. Because most of T is retained near plasma-facing surface or deposited layers making C-H bonds, following three methods have been applied (1) thermal desorption; (2) physically removal of near-surface layers such as mechanical methods, sputtering by inert gas ions, or laser ablation; (3) chemical removal by reactions with reactive gases like H_2O and O_2 or their ions or plasma; and (4) isotopic exchange of T with D.

Although thermal desorption is one of the most reliable ways to remove T from whole volume, required temperature for full removal is quite high. Because of thermal stress caused by heating, simple heating of plasma-facing tiles in tokamak to very high temperature is not realistic. Although laser irradiation works as surface heating for thermal desorption of surface recession by ablation, its effects are limited near surface layers [44, 45]. He grow discharges have been successfully applied in tokamaks for cleaning of their PFS [46, 47]. Unfortunately, it would not work for removal of T in deposited C layers on plasma shadowed area. Introduction of O_2 gas or O_2 discharge is a good chemical way to remove surface layers including T making volatile molecules of CO, CO_2, and DTO [48]. Nevertheless, remaining O in tokamak and surface oxidation of in-vessel components are concerned for the following discharges, i.e. oxygen impurity in plasma could cause instability of plasma or even disruption. Although isotope exchange of T with H_2O or D_2O is encouraging as observed in JT-60U [49], resultant HTO and DTO are hazardous and recovery of T form them requires large effort. Residual D_2O is also concern for following discharges.

9.7.2 Removal/Recovery of T Retained in W

T removal from graphite is not easy, so as T removal from plasma-facing W. Most of T in W are trapped with larger trapping energy. Therefore, high temperature is required to thermally desorb T in W [20, 50]. Furthermore, depth distribution of T in W would be significantly different from that in C. Because of faster T diffusion in W, T can easily penetrate in deep and be trapped. This means only thermal desorption would work as reliable method to remove T in plasma-facing W.

Efficiency of thermal desorption for tritium retention measurement and removal in ITER was studied [32]. Although isotopic exchange with H or D would work, T in deep inside is quite hard to remove by the isotope exchange [51]. Ogorodnikova, et al. [52] have tried the removal of D trapped at radiation damages in tungsten by isotopic exchange with H atomic beam. Although the efficiency of isotopic exchange increases with increasing the sample temperature, incident ion/atomic flux, and incident H energy, full removal was difficult. Roth et al. [51] have concluded that the isotopic exchange is not a viable method to reduce the tritium inventory in ITER.

Laser ablation is tried to remove H implanted in W. Owing to heating the surrounding area, some hydrogen is thermally released in addition to the ablation. Hence, laser power near ablation threshold is recommended [53]. Deuterium desorption by laser heating was studied by Yu et al. [54], and comparison of laser desorption with the TDS was made by Zlobinski et al. [55].

Finally, it should be mentioned that C would be easier to remove T than W. Although small T retention has been one of the most important criteria for selection of PFM, easy removal of retained T would be more important in the aspect of T safety and fuel self-sufficiency. In both respects, C would be a better PFM candidate than W [11, 20].

9.8 Summary

Accounting or quantitative estimation/determination of T inventory in a reactor is one of the critical issues to establish fuel self-sufficiency and to avoid radio-hazardousness. At present, however, the estimation of T inventory in a reactor includes large error owing to the lack of data of H retention in PFM under the reactor relevant conditions, in particular dependence of the H retention on fluence and temperature for both PFM candidates of C and W remain as urgent tasks.

The mechanism of fuel retention is different between C-wall and W-wall. T corporation with C deposited layers on plasma shadowed area dominates the T retention in the former, while T retention in bulk (by diffusive penetration and trapping) dominates in the latter. For both cases, the total T inventory after the long discharges seems to be comparable level.

To attain the fuel self-sufficiency, the allowable fuel retention rate in PFM should be less than around 0.1% and removal of T from PFM seems dispensable. To keep T safety the maximum T inventory in the reactor is set to be around 1 kg. Therefore, the reduction of the fuel retention rate and the fuel recovery become necessary tasks.

Although it seems quite hard to attain the fuel retention rate of 0.1%, it could be attained after long time discharge or at the steady-state discharge in a reactor. Because after long time plasma exposure, the rather thick surface layers of PFM must be saturated with the fuel and only tiny part of injected fuel would diffuse into deep. Higher temperature operation in a reactor compared to present tokamaks should significantly reduce the saturation concentration. Thus, both the C-wall and the W-wall could attain the fuel retention rate below 0.1% at the steady-state operation. Nevertheless, the total fuel retention in PFM must continuously increase owing to the increase of deposited layers and deep penetration and could come across the site limit given by T regulation low. H trapping in defects produced in PFM would add the total fuel retention, which is still very uncertain. Hence, the removal or recovery of T from PFM is mandatory not only for the fuel self-sufficiency but also for T safety and the development of T removal technique from PFM is urgent. As for the selection of PFM of a reactor, easy fuel removal is an important criterion different from ITER for which it was the less fuel retention.

References

1. T. Tanabe, *Tritium: Fuel for Fusion Reactors*, Springer Japan, 2017 ISBN: 978-4-431-56458-4
2. J. Roth, E. Tsitrone, T. Loarer et al., Tritium inventory in ITER plasma-facing materials and tritium removal procedures. Plasma Phys. Control. Fusion **50**, 103001 (2008)
3. J. Roth, K. Schmid, Hydrogen in tungsten as plasma-facing material. Phys. Scr. **T145**, 014031 (2011)
4. A. Kirschner, D. Borodin et al., Modelling of tritium retention and target lifetime of the ITER divertor using the ERO code. J. Nucl. Mater. **363–365**, 91 (2007)
5. Y. Gotoh, T. Tanabe, Y. Ishimoto et al., Long-term erosion and re-deposition of carbon in the divertor region of JT-60U. J. Nucl. Mater. **357**, 138–146 (2006)

6. T. Tanabe, K. Masaki, K. Sugiyama, M. Yoshida, An overview of recent erosion/deposition and hydrogen retention studies in JT-60U. Physica Scripta **T138**, 014006 (9 pp) (2009)
7. M. Yoshida, Retention mechanism of hydrogen isotopes in plasma facing wall of JT-60U, Doctoral dissertation. Kyushu University (2011)
8. M. Yoshida, T. Tanabe, K. Sugiyama et al., Long term hydrogen retention in the plasma facing carbon tiles of JT-60U. J. Nucl. Mater. **417**, 620–623 (2011)
9. M. Yoshida, T. Tanabe, T. Hayashi, et al. Construction of a fuel retention model for full carbon devices. Physica Scripta **T145**, 014023 (5 pp) (2011)
10. M. Yoshida, T. Tanabe, A. Adachi et al., Carbon transport and fuel retention in JT-60U with high temperature operation based on postmortem analysis. J. Nucl. Mater. **438**, S1261–S1265 (2013)
11. T. Tanabe, Revisiting Carbon Materials as Plasma-Facing Materials of a Fusion Reactor. Plasma Phys. Rep. **45**, 300–314 (2019)
12. H. Atsumi, T. Tanabe, T. Shikama, Hydrogen behavior in carbon and graphite before and after neutron irradiation–Trapping, diffusion and the simulation of bulk retention–. J. Nucl. Mater. **417**, 633 (2011)
13. K. Niwase, T. Tanabe, M. Sugimoto, F.E. Fujita, Laser Raman microprobe analysis of the modified surface of He^+-irradiated graphite. J. Nucl. Mater. **162–164**, 856 (1989)
14. H. Atsumi, M. Iseki, T. Shikama, Trapping and detrapping of hydrogen in carbon-based materials exposed to hydrogen gas. J. Nucl. Mat. **212–215**, 1478 (1994)
15. T. Tanabe, K. Masaki, Y. Gotoh et al., Review of retention of hydrogen isotopes (H, D, T) and carbon erosion/deposition in JT-60U. Atomic Plasma-Mater. Interact. Data Fusion **15**, 73–90 (2011). and references therein
16. N. Miya, T. Tanabe, M. Nishikawa et al., Research activities using plasma facing materials in JT-60U under a cooperative program between JAERI and universities. J. Nucl. Mater. **329–333**, 74–80 (2004)
17. J.P. Coad, N. Bekris et al., Erosion/deposition issues at JET. J. Nucl. Mater. **290–293**, 224–230 (2001)
18. P. Roubin, B. Pégourié, R. Smirnov, Analysis of carbon deposited layer growth processes in Tore Supra. J. Nucl. Mater. **390–391**, 49–52 (2009)
19. R. Behrisch, G. Federici, A. Kukushkin, D. Reiter, Material erosion at the vessel walls of future fusion devices. J. Nucl. Mater. **313–316**, 388–392 (2003)
20. T. Tanabe, Review of hydrogen retention in tungsten. Physica Scripta **T159**, 014044 (12 pp) (2014)
21. Y. Hatano, M. Shimada, T. Otsuka, Deuterium trapping at defects created with neutron and ion irradiations in tungsten. Nucl. Fusion **53**, 073006 (2013)
22. H.S. Zhou, L.P. Zhao K. Umstadter, et al. Exposure of vacuum plasma spraying tungsten to PISCES and DIII-D plasmas. Fusion. Eng. Design **86**, 325 (2011)
23. A. Rusinov, M. Sakamoto, H. Zushi, Deuterium retention in plasma sprayed tungsten exposed to low energy deuterium plasma. Plasma Fusion **8**, 1405004 (2013)
24. T. Otsuka, T. Tanabe and K. Tokunaga, behavior of tritium in plasma-sprayed tungsten coating on steel exposed to tritium plasma. Physica Scripta **T145**, 014035 (4 pp) (2011)
25. Z. Tian, J. Davis, A.A. Haasz, Deuterium retention in tungsten at fluences of up to 10^{26} D^+/m^2 using D^+ ion beams. J. Nucl. Mater. **399**, 101 (2010)
26. K. Sugiyama, M. Mayer, V. Rohde et al., Deuterium retention in bulk tungsten exposed to the outer divertor plasma of ASDEX Upgrade. Phyica Scripta **T145**, 014033 (2011)
27. N. Enomoto, S. Muto, T. Tanabe, Grazing-incidence electron microscopy of surface blisters in single- and polycrystalline tungsten formed by H^+, D^+ and He^+ irradiation. J. Nucl. Mater. **385**, 606–614 (2009)
28. M.H. J 't Hoen, B. Tyburska-Puschel, K. Ertl et al., Saturation of deuterium retention in self-damaged tungsten exposed to high-flux plasmas. Nuclear Fusion **52**, 023008 (2012)
29. O.V. Ogorodnikova, J. Roth, M. Mayer, Ion-driven deuterium retention in tungsten. J. Appl. Phys. **103**, 034902 (2008)

30. VKh Alimov, K. Ertl, J. Roth, Deuterium retention and lattice damage in tungsten irradiated with D ions. J. Nucl. Mater. **290–293**, 389–393 (2001)
31. N. Bekris, C. Caldwell-Nichols, L. Doerr et al., Possible techniques for the detritiation of first wall materials from fusion machines. J. Nucl. Mater. **307–311**, 1649–1654 (2002)
32. J. Roth, E. Tsitrone, T. Loarer, Tritium inventory in ITER plasma-facing materials and tritium removal proceduresPlasma Phys. Control. Fusion **50**, 103001 (2008)
33. M.J. Gouge, Fueling of ITER-scale fusion plasmas. Fusion Technol. **28**, 1644 (1995)
34. P. Andrew, P.D. Brennan, J.P. Coad, Tritium retention and clean-up in JET, 1999 Fusion Eng. Des. **47**, 233–245 (1999)
35. W. Jacob, C. Hopf, M. Schlüter, Chemical sputtering of carbon by combined exposure to nitrogen ions and atomic hydrogen. Phys. Scr. T. **124**, 32–36 (2006)
36. F.L. Tabar´es, D Tafalla, I Tanarro, et al, Suppression of hydrogenated carbon film deposition by scavenger techniques and their application to the tritium inventory control of fusion devices. Plasma Phys. Control. Fusion **44**, L37 (2002)
37. F.L.Tabar´es, V. Rohde, Suppression of carbon re-deposition by nitrogen injection in the sub-divertor region of ASDEX-Upgrade. Nuclear Fusion **45**, L27 (2005)
38. W. Bohmeyer, F.L. Tabares, M. Baudach et al., The scavenger effect—Does it work? J. Nucl. Mater. **390–391**, 560–563 (2009)
39. C.H. Skinner, N. Bekris, J.P. Coad et al., Tritium removal from JET and TFTR tiles by a scanning laser. J. Nucl. Mater. **313–316**, 496–500 (2003)
40. C. Grisolia, A. Semerok, J.M. Weulersse et al., In-situ tokamak laser applications for detritiation and co-deposited layers studies. J. Nucl. Mater. **363–365**, 1138 (2007)
41. N. Bekris, J.P. Coad, A. Widdowson et al., Assessment of the flash-lamp photon-cleaning detritiation method tested at JET. J. Nucl. Mater. **390–391**, 614–617 (2000)
42. C. Grisolia, S. Rosanvallon, P. Coad et al., JET contributions to ITER technology issues. Fusion Eng. Des. **81**, 149–154 (2005)
43. C.H. Skinner, J.P. Coad, G. Federici et al., Tritium removal from carbon plasma facing components. Phys. Scr. **T111**, 92–97 (2004)
44. Y. Sakawa, K. Satoh, T. Shibahara, T. Tanabe, Ablation of hydrogen-implanted graphite target using pulsed laser beam. Fusion Eng. Design **81**, 381–384 (2006)
45. C.H. Skinner, N. Bekris, J.P. Coad et al., Tritium removal from JET and TFTR tiles by a scanning laser. J. Nucle. Mater. **313–316**, 496–500 (2003)
46. D. Mueller, W. Blanchard, J. Collins et al., Tritium removal from TFTR. J. Nucl. Mater. **241–243**, 897–901 (1997)
47. K. Itami. N. Asakura. H. Tamai et al., RF heated wall conditioning discharges in JT-60U. J. Nucl. Mater. **390–391**, 983–987 (2009)
48. V. Philipps, G. Sergienko, A. Lyssoivan et al., Removal of carbon layers by oxygen glow discharges in TEXTOR. J. Nucl. Mater. **363–365**, 923–929 (2007)
49. M. Yoshida, T. Tanabe, K. Sugiyama et al., Characterization of in-vessel hydrogen isotope retention of JT-60U. J. Nucl. Mater. **415**, S752–S756 (2011)
50. R.A. Causey, J.N. Brooks, G. Federici, Tritium inventory and recovery in next-step fusion devices. Fusion Eng. Design **61–62**, 525–536 (2002)
51. J. Roth, T. Schwarz-Selinger, VKh Alimov, E. Markina, Hydrogen isotope exchange in tungsten: Discussion as removal method for tritium. J. Nucl. Mater. **432**, 341–347 (2013)
52. O.V. Ogorodnikova, S. Markelj, V.S. Efimov et al., Deuterium removal from radiation damage in tungsten by isotopic exchange with hydrogen atomic beam. J. Phys: Conf. Ser. **748**, 012007 (2016)
53. K. Hirata, K. Furumoto, N. Yamamoto, T. Tanabe, Application of pulsed laser irradiation for removal of hydrogen retained in tungsten. J. Nucl. Mater. **443**, 298–301 (2013)
54. J.H. Yu, M. Simmonds, M.J. Baldwin, R.P. Doerner, Deuterium desorption from tungsten using laser heating. Nuclear Mater. Energy **12**, 749–754 (2017)
55. M. Zlobinski, V. Philipps, B. Schweer et al., Hydrogen retention in tungsten materials studied by Laser Induced Desorption. J. Nucl. Mater. **438**, S1155–S1159 (2013)

Chapter 10
Selection of Plasma-Facing Materials

10.1 Criteria for Selection of PFM

In a fusion reactor, the combination of very high-power load and resulting high temperatures, high radiation levels, intense production of transmuting elements and high thermomechanical loads requires very high-performance for plasma-facing materials (PFM). Erosion of plasma-facing components (PFC) determines their lifetime and generates a source of impurities, which cool down and dilute the plasma, while deposition causes T retention. The resuspension of dust could be a consequence of loss-of-coolant accidents (LOCA) and lss-of-vacuum accidents (LOVA), and it can be dangerous because of its radioactivity, toxicity, and capability causing a hydrogen explosion [1].

Accordingly, the selection of PFM for a fusion reactor is a crucial issue. As described in Chap. 1, PFM in plasma apparatus has been changing with improvement of plasma confinement, targeting higher density, and temperature for self-burning. Now major criteria for the selection of PFM are (1) tolerance to high-power load, (2) long lifetime including low erosion and tolerance to neutron damage, and (3) low tritium retention. To be tolerant to the first one, high melting point, high thermal conductivity, and high heat shock resistance are mandatory. At present, mainly W and carbon-based materials are considered to be PFM of the reactor. Although liquid wall using liquid metals like Li is attractive and could be alternative, significant effort is required to realize the liquid wall with reliability.

There are opinions on pros and cons of the selection of PFM in various detailed criteria, such as (1) low radiation in plasma or low impurity release to plasma, (2) surface damage and material loss, (3) wall life time, (4) tritium retention and easy recovery, (5) neutron irradiation damage including activation, (6) maintenance, and so on. In Table 10.1, important characters relating plasma materials interactions (PMI) are compared for three PFM candidates, C, Be (Beryllium), and W. Low Z materials like C and Be are much better than W in the aspect of plasma contamination, while material loss by erosion is much larger for the former than the latter. The radiation loss by impurities in the plasma center depends on their concentration, and

T. Tanabe, *Plasma-Material Interactions in a Controlled Fusion Reactor*, Springer Series in Plasma Science and Technology, https://doi.org/10.1007/978-981-16-0328-0_10

Table 10.1 Summary of behavior of Carbon material, Beryllium, and High Z as PFM armor

	Carbon-based material	Be	High Z
General	Much experience in large tokamaks, JET, JT-60U, TFTR, ASDEX, DIII-D TEXTOR, Tore Supra, and others	Experience only in JET	Some tokamak experience in JET, ASDEX and EEST Dedicated experiments in TEXTOR and ALCATOR
Radiation Allowable level in burning plasma	Low 10^{-4}	Low 10^{-3}	High 10^{-6}
Thermal response	No melting allows highest Temp. use Low thermal conductivity at high Temp.	Low melting Temp. limits <1000 K	High melting Temp. Crystalline growth limits maximum Temp. <1600 K Embrittlement limits low Temp. >800 K Thermal fatigue limits long time use
Erosion	High Low physical sputtering threshold High chemical erosion RES given by impurity ions limits maximum Temp. <1600 K	High physical/chemical sputtering Evaporation	Low High physical sputtering threshold Sputter erosion by impurity Melting/crack erosion
Deposition	Large on plasma shadowed area	Large on plasma shadowed area	Prompt deposition
He effect	Little	Blistering	Blister erosion
Tritium retention	Large in deposited layers Significant reduction above 800 K	Significant trapping in defects	T trapping near surface T dissolution in bulk
Neutron damage	Low activation. Degradation of thermal conductivity Swelling	Low activation He transmutation	Medium activation Embrittlement and DBTT in crease Transmutation to Re and Os

maximum allowable concentration is around 10^{-3} and 10^{-6} for C and W, respectively. However, it should be noted that some impurities are required for edge cooling or power exhaust and impurity seeding in which low Z to medium Z materials like N, Ne, and Ar are intentionally introduced in the boundary plasma. In this respect, C-wall has advantage because eroded C works as a radiator without additional impurity seeding.

Although high radiation of high Z materials accumulated in plasma center is serious concern, their high radiation in the boundary plasma is attractive for cooling if they were localized in the boundary plasma. In recent tokamak experiments using

W as PFM, it seems possible to avoid the W accumulation in the plasma center with appropriate plasma control. However, operational temperature window for W used as PFM is limited because of quite low allowable impurity level and avoidance of material damage caused by melting and embrittlement. Therefore, the tolerant power load to W would be less than that for C. For C, in spite of its large erosion, damages like cracking and fracture could be avoided by using carbon fiber enforced carbon materials (CFC). Benefit of C is that erosion could be repaired by C deposition by CVD and PCVD processes for example. Low erosion of W is advantage for materials lifetime. However, heat shock damage and materials degradation, i.e. loss of ductility or increased of DBTT by neutron irradiation and recrystallization of W due to cyclic power load are concerned. Low neutron activation of C is another advantage making maintenance and repairing easier.

Since the shape of plasma-facing surface (PFS) could not be ideally smooth both in toroidal and poloidal directions, high heat load at the edges of plasma-facing tiles cannot be avoided. Self-shaping, i.e. erosion at salient region and deposition at concave region, makes PFS flatten and consequently erosion would be reduced. High erosion and deposition of C-wall could be compensated by the self-shaping. Still depositions at plasma shadowed and remote areas are concerns for C. In this respect, W is better owing to less erosion.

Tritium retention is one of the critical issues. Low retention is better and has been one of the most important criteria for the selection of PFM of ITER. However, as described in the previous chapter, easy recovery would be more important for T fuel self-sufficiency. For C-wall, most of T is retained in deposited layers and near-surface regions of eroded area, while bulk retention is dominant for W-wall. In any PFM, T recovery must be done, although required techniques are different.

In summary, W seems to be a very good PFM, if the central accumulation of high Z impurities could be avoided by suitable transport control. "Prompt deposition" of high Z atoms is very promising because it reduces the erosion and long-range transport, consequently deposition at remote area is suppressed. Under the same power load condition, the net power load is likely less for high Z materials than that for low Z materials because of their higher reflectivity both of energy and particle, which must be confirmed in the future.

10.2 Concerns on W Usage as PFM

W seems a good candidate PFM in aspects of low erosion (and hence low deposition) and low T retention. However, various concerns in actual use of W as PFM of burning plasma remain as described in this section. ITER could be one of the most important test beds for W usage as PFM. In particular, it would be one of the most important tasks to study the effects of W on the operation of burning plasma. Unfortunately, however, concerns on materials aspects in long term usage could not be tested in ITER and requires separate examination to realize W PFM in a reactor. Power load tests using linear plasma machines would help this.

A concern on W from the PMI aspect is its accumulation in plasma center resulting in huge radiation loss which could cause plasma claps. In recent experience with using W as PFM in ASDEX, EEST, and JET, the central accumulation could be avoided by appropriate plasma control. However, the power load was kept rather low to avoid material damage. Accordingly, the material's response of W to the high-power load in burning plasma remains quite uncertain. Unclear direct responses to the high-power load are sublimation, melting, movements of melt layer and droplets, re-solidification, and recrystallization. Reponses to repetitive power loads are various. They are loss of ductility due to grain growth or recrystallization, surface modification by He injection like blistering, swelling and fatigue, erosion by melting, and particle release by cracking of recrystallized grains. Neutron irradiation gives radiation damages, activation and nuclear transformation to produce Re and Os of which alloying in W enhances brittleness. Together with cyclic heat load swelling and fatigue limit the lifetime of PFM W. High chemical activity with hot water produces volatile molecules of WO_3 in LOCA which is very dangerous owing to high activation of W by neutron irradiation. These characters do not allow W to use as heat sink, or direct contact to water. Radioactivity of released droplets produced by melting and cracking peace from W surface requires particular care on their safe handling in addition to their T retention. High D and T retention could cause hydrogen embrittlement, in particular, their trapping at grain boundaries is concerned but not studied yet.

Generally, W used as armor tiles should be welded or brazed to heat sink or cooling channel materials, which is not easy because of large difference in thermal expansion coefficients of W and the heat sink materials. It should also be mentioned that W is heavy, nearly ten times compared to C materials and is very hard to machine. This clearly appears as difficulty of manufacturing of the ITER divertor. Thus, significant efforts are still required to design and manufacture W divertor mainly in materials aspect.

10.3 Use of Carbon Materials as PFM

10.3.1 Character of C as PFM

Carbon materials (C) can be very good armor tiles of PFW, if they are used at higher temperatures. Considering the temperature of cooling water for power generation in DEMO [2], around 700 K or higher, the temperature of heat sink is likely above 800 K. Because of their porous nature, the C armor tiles must be adhered to heat sink metals having cooling channels or connected to cooling systems. For adhesion, both brazing and mechanical fixing would be possible, while welding is not available because of carbide formation or large difference in thermal expansion between C and the heat sink metals resulting easy rupture of the bonding. Although the brazing would be better for heat removal, mechanical fixing is recommended for easy exchange of the armor tiles by remote handling. The poorer heat transfer by the mechanical fixing than that by the brazing must result in higher surface temperature. Moreover, poor thermal conductivity of C at higher temperatures would enhance surface temperature rise. One of the benefits of C is non-melting, if significant temperature rise of plasma-facing surface is allowed. In a reactor, periodical breaks for reactor maintenance, for example, at every one or two years, are required and defective tiles can be replaced during the maintenance. This also keeps the neutron irradiation level rather low.

One of the benefits of C is their small hydrogen diffusion in crystallites, which inhibits hydrogen penetration into the crystallites, even though C is porous, resulting quite small bulk retention. In addition, chemical erosion could be significantly reduced, if the temperature of the C PFC was above ~ 800 K, near reactor operating temperature. Still radiation enhanced sublimation at higher temperatures is concerned Less erosion would result in less redeposition. Moreover, the deposited carbon layers at such high temperatures are rather graphitized and their T up-take becomes significantly less. Consequently, hydrogen retention in C seems more easily saturated compared to those for any metals in which no saturation seems to occur owing to hydrogen diffusion in deep inside and even permeation.

It is well known that neutron irradiation degrades thermal conductivity of C and gives dimensional change. However, their high-temperature usage would mitigate the degradation. In particular, their usage as armor tiles allows temperature rise and even the dimensional change. Furthermore, the temperature increase due to the loss of thermal conductivity by neutron irradiation would reduce tritium retention and anneal the damage.

10.3.2 Possible Use of C as PFM in a Reactor

As described above, C can be armor tiles of PFC in a fusion reactor with temperature of cooling water or gas being above 700 K. Mechanical fixing of the C armor tiles to heat sink or cooling channels is recommended allowing surface temperature rise to reduce chemical erosion, T retention, and neutron irradiation effects. Nevertheless, the degradation of thermal conductivity and some dimensional change of C by neutron irradiation are unavoidable, which requires replacement of the tiles. The mechanical fixing ensures easy replacement of failed tiles by remote handling, and periodic break for system maintenance gives opportunity for routine replacements of the tiles. Still concerns remain; how to fix the carbon armor tiles to PFC mechanically, tolerance of the mechanical fixing to surface temperature escalation due to its poor thermal contact and any effects caused by irradiation. It should be mentioned that less activation of C by neutron irradiation than any other candidate plasma-facing metals makes system maintenance safer in aspect of radiation exposure of workers.

In order to reduce T inventory in a reactor, additionally proposed are routine T removal by isotopic exchange with D discharges and installation of replaceable cooled plates at divertor region to acquire carbon deposition and T included in.

In conclusion, carbon materials can be a good PFM in a fusion reactor, if they are used as armor tiles mechanically fixed to heat sink allowing possible temperature escalation of tile surfaces. Based on these considerations, Table 10.2 is summarized how to mitigate concerns on C usage as PFM. Chemical erosion will be significantly reduced owing to chemical instability of C–H bond at higher temperature and consequently depresses the redeposition. Plasma shaping by erosion and prompt deposition will improve the tile alignment and hence reduce the net erosion. Nevertheless, the deposition at gaps of eroded tiles and plasma shadowed area in divertor region cannot be avoided. In case of large erosion at specified location, in situ repairing by carbon deposition using CVD, PVD or plasma-assisted decomposition of methane is possible, which is one of the most important benefits of C.

Degradation of material performance of C by neutron irradiation, in particular, dimensional change and reduction of thermal conductivity, is unavoidable. The mechanical fixing promises easy replacement by remote handling. Less activation by neutron irradiation of C than any other PFC candidate metals helps the system maintenance.

Considering all points for mitigation of concerns for using C as PFM in Table 10.2, the concept of utilization of CFC as PFM is given in Fig. 10.1. Considering high heat load to PFM in a fusion reactor and porous nature of carbon materials, utilization of CFC as PFM is limited as armor tiles. Even so operational window of the tiles is limited. To remove some 10 MW m^{-2} at highest power load area with rather low thermal conductivity of CFC at higher temperatures, the thickness of CFC should be thin enough to ensure the heat removal with thermal contact to a heat sink material. The thermal conductivity of CFC at lower temperature is very high, while it

Table 10.2 Concerns on Carbon materials as PFM armor and their mitigation

Concerns	Remarks	Mitigation/Prospects	
High erosion	Erosion will not continue at same place	In situ repairing by CVD, PVD or plasma-assisted processes	
	Fine alignment of tiles is required	Plasma shaping will also work	
T retention in tiles	Formation of T saturated layers is limited near surface	Saturated concentration (T/C) significantly decreases with Temp. Isotopic exchange works	
T retention in redeposited layers	Location of redeposited layers is predictable	Tile gap: Tile replacement will reduce	
		Divertor: Installation of cooling plate to take-up redeposition	
Dust formation	Mainly caused by exfoliation of redeposited layers	Chemical instability of C-H bond at higher Temp. reduces formation of redeposited layers	
Neutron damage	Loss of thermal conductivity	High Temp. operation decelerates damage formation	Tile replacement
	Dimensional change		
	Increase of T trapping		Isotopic exchange

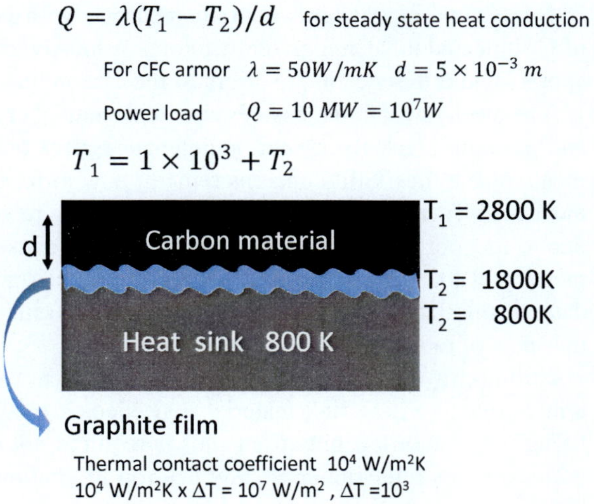

Fig. 10.1 Conceptual scheme for utilization of CFC as PFM armor in a reactor

significantly decreases with temperature and neutron irradiation. For temperatures above 1500 K, reduction of thermal conductivity by neutron irradiation is relieved by simultaneous annealing effect. According to reference [3], we can assume the thermal conductivity of CFC above 1500 K to be around 50 W m^{-1} K^{-1}. Considering an armor tile made of CFC with thickness of 5×10^{-3} m, temperature difference between the front surface and the back surface should be 1000 K to allow power flow of 10 MW m^{-2} as shown in Fig. 10.1. It should be noted that to enhance the thermal contact, thin graphite film is inserted between the carbon armor and the heat think material in the figure. Since the thermal contact coefficient of CFC mechanically fixed to heat the sink material kept at 800 K is not known, we can assume around 10^4 W m^{-2} K^{-1} often observed in metal-metal contact. Taking temperature difference between the CFC and the heat sink to be 1000 K to allow the power flow of 10 M W m-^{2}, the back-side surface would be 1800 K, and consequently the front surface be 2800 K. If the power load is 10 MW m^{-2} for first wall, the temperature difference would be 500 K, which could be tolerable. Radiation cooling would also mitigate the surface temperature rise. At such high temperatures, hydrogen retention must be quite small. Although chemical erosion above 1500 K is not concerned, radiation enhanced sublimation is. Nevertheless, erosion and prompt deposition significantly reduces the net erosion as already discussed. It is somewhat ironical that impurity seeding to enhance radiation cooling is planned in ITER and a reactor, different from development of plasma confinement by reducing impurities in plasma. In some sense, carbon is good radiator in boundary plasma, as appeared carbon blooms in JET [4] and strong carbon radiation at private flux region (MARF) in JT-60U [5], which were not good for plasma confinement at that time. Radiation enhanced sublimation of carbon would enhance radiation and could work as vapor shielding to protect from further erosion. This kind of phenomena can be tested only in ITER.

Benefit of C is possibility of in situ repairing, with using CVD process. The surface of C subjected to such high temperatures is mostly graphitized and more or less no appreciable difference in the origin of the carbon materials.

The mechanical fixing ensures easy replacement of failed tiles by remote handling and periodic break for system maintenance gives opportunity for routine replacements of the tiles. Still concerns remain; how to fix the carbon armor tiles to PFC mechanically, tolerance of the mechanical fixing to surface temperature escalation due to its poor thermal contact, and any effects caused by irradiation. It should be mentioned that less activation of C by neutron irradiation than any other candidate plasma-facing metals (except liquid metals) makes system maintenance safer in aspect of radiation exposure of workers.

Still improvement of thermal contact coefficient for the mechanical fixing of the armor tiles to a heat sink material is needed. It totally depends on the method of fixing. Utilization of thin armor tiles is required not only for thermal conductance but also stress relieving caused by neutron irradiation dimensional change. Higher temperature operation is favored owing to significantly reduction of the neutron irradiation effects although data on neutron irradiation of CFC at elevated temperature is scarce.

10.4 Liquid PFM

Lithium-based liquid surface as a candidate PFM has gained deserved attention in the last decade [6–8]. Table 10.3 compares solid and liquid as PFM [8]. Notable advantages have resulted from the use of Li in various applications including thin-film coatings on metallic substrates, lithium-graphite surfaces, Li based alloys, and others. There are two main challenges to the use of lithium-based surfaces as viable PFM. One is their narrow operating temperature window. At temperatures just above the melting point of Li (e.g. 180°C), non-linear erosion ensues and sputtering increases toward unity. The eroded Li atoms by sputtering are mostly ions and the remaining neutrals quickly ionized due to their low ionization potential. This fact leads to effective screening of eroded Li material from the core plasma at all temperatures. For liquid Li, therefore, macroscopic emission of material due to off-normal events such as ELMs or similar events can be problematic and thus the second challenge using Li as PFS [6]. Still difficulty remains in establishing a liquid cooling system including hazardous radioactivity and unstable chemical nature of Li in atmosphere.

Table 10.3 Advantages and Disadvantages of Solid PFCs and Liquid PFCs [8]

	Solid (W)	Liquid Metal
Lifetime of plasma facing components	PFCs need change-outs	Liquid can be replenished
Max. steady-state power flux	5–10 MW m^{-2}	~30 MW m^{-2} or higher
Resilience to transients (ELMs, disruptions)	**Crack formation**	**Thinning of liquid layer**
Tritium retention	Most likely acceptable	**Huge T uptake in Li**
Low plasma core contamination (Z_{eff} or radiation)	$<10^{-5}$ demonstrated	**Not demonstrated**
Chemical compatibility in PFC system	None for He; known for water	**Critical issue**
Embrittlement of structural materials	**Irradiation embrittlement**	**LM embrittlement**
Radioactivity at end of plant lifetime	**Moderate to high**	Low
Accident tolerance (e.g. leaks, air ingress)	Oxidization, little volatilization	Li is potentially problematic
Maintenance and operation	Concerns are brittleness, melting, and dust	New
Unknown risks	Low	**High**

10.5 Consideration of T Fuel on the Selection of PFM in a Reactor

The in-vessel T inventory must be kept as small as possible, not only for safety reason but also for fuel self-sufficiency. In the present tokamaks using the C-wall, fuel retention rates in their vacuum vessel are significantly large, i.e. a few % of fueled hydrogen is continuously piled up [9, 10]. As described in the previous chapter, the retention rate should be less than 0.1%. Otherwise, DT reactors cannot be an economical energy source. Of course, it is believed that ITER or any reactors should be operated under the steady-state condition, and the in-vessel T inventory must be saturated suggesting the retention rate at very high fluence should be nearly 0 or most of injected fuels are reemitted. Unfortunately, however, the current estimation shows no saturation for any materials combinations of Be, C and W, owing to continuous hydrogen piling up in the deposited layers formed at plasma shadowed and remote areas and also in the bulk. The removal of T in the deposited layers at the plasma shadowed area is not easy and methods for reduction and recovery (removal) of the in-vessel T inventory are still under development and need significant efforts as described in Chap. 9.

From concerns of large T retention in C, ITER decided not to use C as PFM. In the present ITER design, materials selected are Be for first wall, W for divertor. However, utilization of W bulk below their DBTT could result in the total failure through the cracking from W to cooling tubes. Hence, we should keep C materials as an alternative for PFM in a reactor or even in ITER [11, 12].

According to the present estimation of the in-vessel T inventory, it seems impossible to keep it below 1 kg, ITER safety limit, after certain period of operation. Hence, periodic removal or reduction of the in-vessel T inventory will be indispensable. This means that the total T retention might not be a good figure of merit for the selection of PFM in a reactor. Instead, readiness to remove T once retained should be taken into account. Carbon, at ITER operation temperature, would retain larger amount of T compared with Be and W, while T retention is limited in surface and subsurface regions including deposited layers and inner surfaces facing open pores, and in-grain or bulk retention is quite low. Accordingly, T retained in C is rather easily replaced by hydrogen isotopes subsequently impinging by isotopic exchange. Even T in the deposited layers could be replaced isotopically, if their temperature was above 800 K.

In the aspect of T management, C if used above 800 K (in a reactor) seems better than W. Heavy neutron activation of W would result in much dangerous dust than carbon. Dimensional change of carbon by neutron irradiation was claimed, but intentional and periodic replacement in the occasion of reactor opening for maintenance would relieve the problem. As such, figure of merit for the selection of PFM of DEMO reactor might be reconsidered [11].

10.6 Summary

In a fusion reactor, power load (given by radiation and particle fluxes) to PFS from burning plasma must be extraordinarily high such like the surface of a rocket running in the sun is exposed to. For any materials, there is the maximum tolerable power load and the reactor should be designed not to give higher power load than the maximum. Until now, however, such high-power load to large area like divertor has been hardly realized to test the PFM candidate materials. Therefore, extrapolation of current experience and observations in PSI to reactor condition is indispensable. Nevertheless, expected particle flux to the wall of 10^{24} m^{-2} is too large to make reliable extrapolation. Recently, linear plasma machines giving extremely high power are available and various heat load tests are under way for the development of the ITER divertor. However, details of power load given to the divertor area are still uncertain, i.e. energy spectra of incident ions, electrons, and photons are still quite uncertain. One example is vapor shielding. At present, only ITER can be used as a testbed for PMI in a rector. Hence, it is desired to have a chance to change PFM materials from W to other candidate materials, in particular, carbon-based materials even in ITER.

Still operational conditions of ITER are somewhat different from that of a reactor. Simultaneous establishments of production of electricity, T breeding, and keeping nuclear safety under economically acceptable efficiency should be attained. Owing to shortage in T resource, satisfying T fuel self-sufficiency and keeping T safety following T regulation law are must and require more details in PMI. In addition, engineering to establishment of production of electricity keeping efficient cooling of PFM is the main task following ITER considering effects of neutron irradiation. It should be mentioned that reactor operational temperature is even higher than that of ITER. PMI at higher temperature could be significantly different from that observed in present large tokamaks.

Although carbon materials are excluded to use as PFM in ITER, we should not exclude the possibility of C as PFM in a fusion reactor. C is one of the best high-temperature materials with a low activation property. Since the operating temperature of a fusion reactor is likely to be above 500–800 K, we have to re-estimate erosion/deposition and tritium retention at such high temperature. In the reactor, T retention in C is not likely the problem, and erosion could be repaired by deposition of carbon layers between shots and the deposited layers must be graphitized by a succeeding shot. We should continue R&D of PFM including C.

References

1. A. Malizia, L.A. Poggi, J.F. Ciparisse et al., A review of dangerous dust in fusion reactors: from its creation to Its Resuspension in Case of LOCA and LOVA. Energies **9**, 578 (2016)
2. K. Tobita, S. Nishio, M. Enoeda et al., Compact DEMO, SlimCS: design progress and issues. Nuclear Fusion **49**, 075029(10 pp) (2009)
3. V. Barabash, M. Akiba, I. Mazul et al., Selection, development and characterisation of plasma facing materials for ITER. J. Nucl. Mater. **233–237**, 718–723 (1996)
4. The JET Team, The TFTR Team, M. Ulrickson A review of carbon blooms on JET and TFTR, J. Nucl. Mater., **176–177,** 44–50 (1990)
5. H. Kubo, S. Sakurai, N. Asakura et al., Atomic and molecular processes for heat and particle control in tokamaks. Nucl. Fusion **41**, 227–233 (2002)
6. J.P. Allain, C.N. Taylor, Lithium-based surfaces controlling fusion plasma behavior at the plasma-material interface. Phys. Plasmas **19**, 056126 (2012)
7. R.E. Nygren, F.L. Tabarés, Liquid surfaces for fusion plasma facing components—a critical review. Part I: Phys. PSI Nucl. Mater. Energy **9**, 6–21 (2016)
8. J. Rapp, The challenges of plasma material interactions in nuclear fusion devices and potential solutions. Fusion Sci. Technol. **72**, 211–221 (2017)
9. T. Tanabe, N. Bekris, P. Coad et al., Tritium retention of plasma facing component in tokamak. J. Nucl. Mater. **313–316**, 478–490 (2003)
10. G.F. Matthews, J.E.T.E. Contributors, Plasma operation with an all metal first-wall: comparison of an ITER-like wall with a carbon wall in JET. J. Nucl. Mater. **438**, S2–S10 (2013)
11. T. Tanabe, Tritium fuel cycle in ITER and DEMO: issues in handling large amount of fuel. J. Nucl. Mater. **438**, S19–S26 (2013)
12. T. Tanabe, On the possibility of ITER starting with full carbon. Fusion Eng. Design **81**, 139–147 (2006)

Chapter 11
Closing Remarks

The main purpose of this book is to show that plasma-facing surfaces (PFS) respond actively to the fusion plasma and that without understanding what is really happening at PFS, the influence of PFS on plasma cannot be understood. PFS has been considered as passive to the plasma so that changes of plasma-facing material (PFM) would not have a significant influence on plasma performance. Recent plasma material interactions (PMI) studies, however, indicate that plasma performance can be significantly affected when PFM is altered or modified by plasma power and particle load.

Most of our knowledge about plasma confinement for fusion burning has been obtained from plasma machines with a carbon (C) wall as PFM. Due to concerns about T inventory and neutron effects, the C-wall has been changed to a high Z wall and recent successes of the high Z wall in several tokamaks, like ASDEX-upgrade, EEST, and JET with ITER like walls encourage the use of the high Z wall in a reactor.

However, in JET with ITER-like-wall experiments, the plasma character or properties, in particular, edge properties and hydrogen recycling behavior have clearly changed from those observed with full carbon wall. Compared with current tokamaks, the incident flux of energetic photons (radiation) and particles to PFS in a reactor is so large that the current knowledge on boundary plasma and hydrogen recycling may not be directly useful or applicable for understanding PMI in the reactor. Therefore, it is quite important to know how different PMI will be in a reactor with low Z wall and high Z wall. Due to concerns about W material damage, the power load in JET-ITER-like wall is kept rather low and does not simulate PMI with high Z wall in a reactor. Therefore, main effort to study PMI at reactor-like conditions is given to heat load tests with linear plasma machines or plasma guns. However, as described in this book, the response to power load by radiation and particles are different depending on PFM, accompanying different influence on plasma. Without comparisons of plasma behaviors with low Z and high Z as PFM, for example, correct understanding of PMI might be difficult. Furthermore, the power load to PFM is as high as the surface of a rocket going into the sun is exposed to. Experimental apparatus to give such a

T. Tanabe, *Plasma-Material Interactions in a Controlled Fusion Reactor*, Springer Series in Plasma Science and Technology, https://doi.org/10.1007/978-981-16-0328-0_11

high-power load is hardly available. Therefore, appropriate extrapolation is necessary based on fundamental understanding of physics and chemistry of PMI obtained in the present apparatus.

This book aims to understand and control PMI in a reactor. However, power loads to PFS in today's tokamaks are far less than those in the reactor. Furthermore, the effects of neutron damage of PFM on PMI are difficult to predict. Therefore, most of the book describes fundamental physics and chemistry related to PMI in current tokamaks, which hopefully will help the readers to foresee or predict PMI in a reactor.

Index

A
Ablation, 41, 43, 134, 182
ACAT-DIFFUSE, 128
Active cooling, 6, 136
Alcator C-mod, 136
α particle, 138, 139
Amorphous, 79, 80, 88, 107, 144, 145, 166
Arching, 62, 63, 67, 96
Armor tile, 41, 60, 86, 92, 95, 99, 102, 109, 190–194
ASDEX, 29, 188
Atomic reemission, 64, 122
Auger electron, 61, 62

B
Baffle plate, 171
Balmer line, 23, 55, 97
BeH_2, 58, 59, 92, 109, 141
Beryllium (Be), 5, 12, 58–60, 77, 80, 84, 85, 92, 99, 109, 111, 134, 136, 137, 141, 142, 149, 153, 154, 162, 163, 187, 188, 195, 196
Binary collision, 56
Binder, 164
Blanket, 6–8, 15, 38, 44, 45, 116, 161
Blanket system, 5, 44, 45, 60, 117, 146
Blister, 66, 67, 102, 104–109, 125, 138, 149, 174, 179, 188
Blistering, 104, 108, 129, 179, 188, 190
Blister skin, 104, 105, 107, 108
Blob, 41, 42, 96
Boiler, 39
Bolometer, 25–27
Boron (B), 12
Brazing, 110, 190, 191
Breeding material, 45, 117
Breeding ratio, 5, 116, 117, 146
Bremsstrahlung, 6, 9, 20, 21, 38, 42
Bubble, 66–68, 102, 104, 107, 108, 121, 166, 179
Bulk retention, 156, 163, 168, 172, 178, 189, 191, 196
Burning efficiency, 116, 125, 145–147
Burning plasma, 3, 5, 6, 8–10, 14, 25, 38, 39, 41, 42, 45, 71, 99, 116, 117, 125, 126, 128, 138, 147, 152, 188–190, 196
Burst like emission, 67

C
Calorimeter, 25
Carbon bloom, 45, 68, 99, 101, 102, 193
Carbon Fiber enforced Carbon materials (CFC), 43, 75, 80, 81, 83, 84, 151, 164, 166, 181, 189, 192–194
CH_4, 58, 66, 79, 91
$^{13}CH_4$ injection, 90
Characteristic X-ray, 42, 62
Charge exchange, 6, 41, 86, 118, 141
C-H bond, 59, 164, 165, 181, 192
Chemical potential, 118, 119, 123
Chemical removal, 182
Chemical sputtering, 43, 58–60, 75, 77, 87, 91, 92, 97, 100, 101, 109, 121, 141, 148, 149, 153, 188
Chemical Vapor Deposition (CVD), 176, 189, 192, 194
Closed pore, 164
Collector probe, 23, 28
Confinement, 5, 9–12, 14, 17, 29, 37, 42, 43, 45, 56, 71, 95, 110, 115, 125, 133,

T. Tanabe, *Plasma-Material Interactions in a Controlled Fusion Reactor*, Springer Series in Plasma Science and Technology, https://doi.org/10.1007/978-981-16-0328-0

137, 144, 147, 149, 157, 161, 187, 193, 199
Convexity, 105, 106
Cooling system, 6, 95, 190, 194
Coulomb collision, 9
Cracking, 37, 85, 95, 98, 99, 102, 103, 189, 190, 196
Creep, 100
Crystallite, 119, 153, 164–166, 172, 191
C-wall, 17, 45, 55, 68, 86, 92, 136–138, 140, 141, 144, 145, 148–150, 153, 154, 157, 163, 164, 172, 173, 179, 180, 183, 188, 189, 195, 199
Cyclic heat load, 83, 100, 110, 190

D
D_2O, 64, 182
Dα, 31, 55, 56
DD discharge, 150, 151, 154, 155, 166, 167
DEMO, 39, 178, 190, 196
Density control, 12, 44, 99, 109, 115, 126, 148, 149, 152
Deposited area, 77, 81, 85, 96, 144, 168, 171
Deposition dominated area, 163
Deposition profile, 31, 32, 53, 75, 76, 81, 82, 85–90, 138, 141
Deposition rate, 33, 138, 140, 141, 155, 163, 169–171
Depth profile, 52, 85, 108, 119, 150, 153, 166, 174–176, 178
Desorption, 11, 14, 56, 61, 62, 64, 97, 118, 122, 124, 182
Detached plasma, 23, 24, 45, 70
Detrapping, 44, 118, 119
Detritiation, 180
Deuteron, 6, 168
Diagnostic, 21, 23, 28, 29, 31, 34, 51, 55, 63, 71
DIII-D, 28, 31, 188
Dimensional change, 191, 192, 194, 196
Direct implantation, 172
Direct transport, 79, 90, 91
Discharge cleaning, 12
Displacement, 44
Disruption, 37, 39, 43, 45, 46, 80, 95–99, 133–135, 143, 144, 182, 195
Disruption mitigation, 96, 117, 144
Dissociation energy, 118
Divertor Material Evaluation System (DiMES), 18, 28, 31
Divertor simulation, 97
Divertor target, 6, 42, 43, 45, 63, 81, 99, 127, 133, 138
Dome-shaped blister, 105, 107
Droplet, 85, 92, 97, 103, 143, 145, 190
DTE campaign, 80
DTO, 182
D-T reaction, 7, 38, 42, 57, 66, 138
Ductile to Brittle Transition Temperature (DBTT), 103, 107, 188, 189, 195
Dust, 34, 80, 83, 85, 92, 143–145, 157, 187, 192, 195, 196
Dynamic retention, 121, 122, 124, 153

E
EAST, 21, 136
Easy recovery, 187, 189
Edge cooling, 57, 63, 77, 101, 102, 146, 188
Edge Localized Mode (ELM), 41–43, 63, 95–97, 104, 126, 133–135, 143, 163, 194, 195
Effective plasma charge (Z_{eff}), 101, 137, 195
Electron cyclotron heating, 9
Electron emission, 61, 62
Electron excitation, 8
Electron temperature, 23, 26–28, 101
Embrittlement, 44, 99, 188, 189, 195
Energetic particle, 5, 6, 37, 44, 49, 51, 52, 57, 62, 63, 127, 138, 168
Energy conversion, 5–10
Energy deposition, 10, 14, 53
Energy loss, 9, 10, 43, 51, 52, 62, 107, 119
Energy source, 4, 5, 11, 12, 15, 43, 45, 133, 145, 161, 195
Eroded area, 60, 77, 80, 81, 85, 150, 155, 167, 168, 171, 189
Erosion dominated, 32, 81, 138, 163
Excited atom, 14, 41, 55, 97, 128
Exhaust, 12, 15, 70, 127, 128, 145, 148, 153, 161, 162, 164

F
F-82H, 176
Filler, 164–167, 172
Fission reactor, 5, 6, 39, 44
Flakes, 34, 83, 84
Flaking, 37
Flash-lamp, 181
Fluence dependence, 153, 171, 173, 174, 177–180
Flux dependence, 59, 60, 180
Fracture, 67, 99, 143, 189
Frenkel pair, 52, 119
Fuel balance, 9, 14, 15, 140
Fuel dilution, 149

Fueling, 3, 99, 115, 125, 128, 145, 147, 149, 152, 157, 161, 164
Fueling efficiency, 5, 116, 125, 145–147, 181
Fuel pin, 5, 39
Fuel recovery, 181, 183
Fuel recycling, 3, 15, 17, 42, 44, 110, 126, 137
Fuel removal, 180, 181, 183
Fuel retention rate, 117, 140, 146, 147, 155, 162, 168–171, 179, 180, 183, 195
Fuel self-sufficiency, 3, 5, 6, 15, 45, 71, 115, 145–147, 152, 157, 162, 179–183, 189, 195, 196
Fugacity, 108, 178
Full carbon machine, 163
Full W machine, 164
Fuzz, 66–68, 109

G

γ–ray, 21, 23
Gas Driven Permeation (GDP), 123
Gas pressure, 45, 86, 104, 105, 107, 108
Gas puffing, 45, 79, 99
Gettering, 12, 109
Grain boundary, 102, 105, 106, 174, 175, 179, 190
Graphite, 20, 29, 30, 65, 79, 80, 83, 84, 98, 122, 123, 164, 166, 182, 193, 194
Graphitization, 79, 166
Grow discharge, 115, 182
Gyration, 60, 98, 109

H

H/C, 60, 165, 166
H_2O, 64, 117, 126, 146, 149, 151, 182
Hα, 27, 30, 71
Hardening, 44, 52, 99
3He, 138
Heat influenced zone, 37, 102
Heat shock resistance, 95, 187
Heat sink material, 103, 110, 176, 190, 192, 194
He bubbles, 66, 67
He glow discharge, 115
He pumping, 17
He remission, 67
HH discharge, 150, 151, 155, 166, 167
High-performance, 187
High recycling, 115
High Z material, 11, 29, 77, 99, 136, 188, 189
H-mode, 31, 134, 137
Hot spot, 62, 63
H saturated layer, 104, 107, 108, 121, 174, 178, 179
HTO, 182
Hydrocarbon, 11, 56, 60, 61, 66, 68, 79, 86, 87, 90, 92, 97, 100, 101, 117, 148, 153, 157, 181
Hydrogen embrittlement, 102, 128, 190
Hydrogen explosion, 187
Hydrogen isotope, 126, 149, 155, 178, 196
Hydrogen recycling, 12, 14, 15, 43, 52, 55, 97, 99, 109–111, 115–117, 119, 120, 128, 129, 157, 199
Hydrogen saturated layer, 100, 101, 165
Hydrogen solubility, 92, 102, 109

I

Implanted profile, 119
Impurity seeding, 41, 45, 70, 99, 101, 116, 117, 135, 188, 193
Inert gas, 12, 45, 63, 66, 117, 181
Inertial cooling, 136
Inner divertor, 23, 24, 33, 60, 75, 78, 79, 81, 88, 91, 134, 138, 140, 141, 150, 151, 154, 163, 170, 171, 173
Inner first wall, 6, 42, 75, 138
Inner surface, 81, 108, 166–168, 196
In situ repairing, 192
International Tokamak Engineering Reactor (ITER), 5, 11, 12, 14, 17, 39, 40, 42–46, 51, 55, 60, 64, 70, 92, 95, 96, 99, 102, 103, 109, 110, 115, 121, 122, 124, 126, 129, 133, 134, 136, 137, 141, 142, 147–149, 151, 152, 154–156, 162, 163, 171–173, 178, 180–183, 189, 190, 193, 195–197, 199
Interstitial, 52, 68, 119, 121, 165
Interstitial cluster, 52, 68, 119
Interstitial loop, 52
In-vessel fuel inventory, 148, 156
In-vessel T inventory, 58, 121, 156, 195, 196
Ion cyclotron heating, 9
Ion Driven Permeation (IDP), 123–125
Ionization, 8, 51, 52, 68, 70, 77, 91, 98, 109, 118, 156, 194
Isotope effect, 125, 129, 161, 162
Isotope separation, 145, 146, 161
Isotopic replacement, 126, 127, 149–151, 155, 166, 168, 178
ITER-Like Wall (ILW), 85, 92, 136, 137, 149, 153

J

JET, 12, 21, 22, 31, 33, 34, 43, 45, 55, 59, 68, 80–83, 85, 86, 90, 99, 101, 109, 110, 126, 127, 133–137, 141–143, 149, 153–155, 188, 190, 193, 199
JET ILW, 92, 136, 141
JET ITER-like wall, 84, 109
JT-60U, 12, 21, 23, 24, 28, 34, 45, 68, 70, 75, 76, 78, 79, 81–83, 86–91, 99–101, 126, 133, 138, 140, 141, 143, 144, 150–152, 155, 164, 166–173, 180, 182, 188, 193

K

Kinetic emission, 61

L

Laboratory experiment, 110, 116, 164, 174, 178
Langmuir prove, 23
Lateral stress, 105, 107, 108
Li injection, 21
Limiter, 10, 11, 20, 21, 26–32, 80, 81, 83, 98, 122, 135, 136, 141, 156
Linear plasma machine, 100, 103, 133, 136, 153, 189, 196, 199
Line of site deposition, 86, 155
Liquid metal, 12, 187, 194, 195
Liquid wall, 187
Lithium (Li), 8, 11, 22, 62, 109, 111, 187, 194, 195
Local melting, 62
Long range transport, 58, 91
Loss of Coolant Accidents (LOCA), 145, 187, 190
Loss Of Vacuum Accidents (LOVA), 145, 187
Louvers, 80, 82, 86, 90
Low recycling, 115, 119, 149, 152
Low Z material, 187, 189

M

Maintenance, 34, 181, 187, 189, 191, 192, 194–196
MARFE, 45, 68, 101
Marker tile, 28
Mark-IIA divertor, 153
Mass balance, 75
Material degradation, 37, 39, 44, 68, 96, 100, 189, 192
Material lifetime, 99
Material performance, 95, 192
Materials balance, 154
Materials property, 37, 52, 68, 110
Materials transport, 138
Mechanical fixing, 190–192, 194
Mechanical removal, 140, 181
Melting, 6, 11, 12, 27, 29, 37, 39, 41, 43, 67, 68, 95, 97–99, 102–104, 109, 134, 136, 187–191, 194, 195
Melting damage, 38, 39
Melting temperature, 10–12, 41, 60, 61, 70
Melt layer, 43, 103, 143, 190
Melt layer motion, 37, 97, 103, 104
14 MeV neutron, 6–8, 38, 44, 95, 116
Micro balance, 28, 29
Microstructure, 83, 84, 102
Modeling, 100, 128
Molecular dynamic simulations, 128
Molybdenum (Mo), 11, 26–29, 55, 56, 65, 77, 98, 122, 136
Movable limiter, 28

N

Net erosion, 75, 77, 141, 142, 192, 193
Neutral beam injection, 9
Neutral flow, 86, 88, 90, 91
Neutron activation, 45, 189, 196
Neutron irradiation damage, 187
Neutron multilayer, 5, 60, 117
Nonthermalized fuel particle, 163
Nuclear collision, 7–9, 51, 52, 118, 119
Nuclear graphite, 164
Nuclear Reaction Analysis (NRA), 142, 177

O

O_2 discharge, 182
Off-normal event, 37, 163, 194
OFMC code, 138
Ohmic heating, 9
Open pore, 153, 164, 166, 167, 196
Operational temperature window, 103, 189
Optical emission, 21, 157
Optical spectroscopy, 23, 30
ORMAK plasma, 11
Osmium (Os), 188, 190
Out divertor, 138
Outer first wall, 42, 75, 138, 172
Out gassing, 181

P

Particle flux, 41, 45, 61, 75, 77, 85, 101, 120, 138, 139, 148, 157, 162, 171, 179, 196
Permeation, 29, 108, 110, 118, 120, 121, 123–125, 128, 161, 176, 178, 191
Permeation barrier, 121
Permeation spike, 123–125
Photoelectron, 62, 63
Photon, 6, 8–10, 14, 38, 41, 42, 49, 51, 52, 61–63, 68, 71, 75, 96, 97, 110, 128, 196, 199
Photon emission, 9, 37, 54, 55, 62
Physically removal, 181
Physical sputtering, 56–59, 77, 91, 137, 138, 188
Plasma center, 10, 20, 21, 23, 24, 27, 30, 98, 136, 187–189
Plasma collapse, 39, 42, 61
Plasma contamination, 11, 37, 57, 187
Plasma detachment, 41, 99
Plasma Driven Permeation (PDP), 123
Plasma induced desorption, 61
Plasma shadowed area, 58, 60, 71, 78, 80, 82, 86, 88, 92, 127, 148, 153, 163, 167–172, 180, 182, 183, 188, 192, 195
Plasma shaping, 192
Plastic deformation, 105, 106
Poloidal side, 86, 88
Postmortem analysis, 28, 30, 31
Potential energy diagram, 118
Powder Metallurgy (PM), 98, 105, 174
Power exhaust, 3, 10, 17, 39, 41, 45, 68, 70, 95, 188
Power generation, 6, 7, 141, 146, 190
Power load test, 25, 39, 40, 63, 133, 136, 189
Private flux region, 78, 79, 90, 91, 101, 150, 193
Projected range, 53, 104, 108, 117, 153
Prompt deposition, 57, 75, 77, 78, 85–87, 91, 92, 101, 154, 155, 188, 189, 192, 193
Proton Induced X-ray Emission (PIXE), 62
Pumping, 75, 79, 88, 143, 151, 155, 181
Pumping duct, 60, 78, 80, 82, 86, 90, 92
Pumping speed, 45, 162

Q

Quartz Micro Balance (QMB), 31, 33

R

Radial distribution, 125, 147
Radiation, 3, 5, 6, 9–12, 15, 17, 20, 21, 23–25, 27–29, 38, 41–45, 49, 61–63, 68, 70, 95, 97, 98, 100, 101, 110, 129, 134, 136–138, 187, 188, 190, 191, 193–196, 199
Radiation cooling, 41, 193
Radiation damage, 14, 42, 49, 52, 68, 95, 109, 182, 190
Radiation Enhanced Sublimation (RES), 58, 59, 61, 77, 100, 101, 188, 191, 193
Radiative cooling, 117
Ramp-down, 42
Ramp-up, 42, 61, 66
Recoil atom, 8
Recombination, 64, 68, 70, 86, 121, 123, 124
Recombinational release, 121
Recombination plasma, 68
Recrystallization, 37, 39, 98, 99, 102, 103, 189, 190
Recycling coefficient, 97, 117, 148, 152
Redeposit, 89
Redeposited layer, 82, 90, 148, 156, 192
Reemission, 12, 37, 51, 53, 63–70, 104, 115, 117, 118, 120, 121, 123, 129, 149, 152, 153
Refinement, 115, 145, 161
Reflection, 29, 37, 49, 51, 53, 54, 62–64, 97, 118, 120, 129, 148, 152, 180
Reflection coefficient, 30, 54, 55, 63, 66, 121
Refueling, 15, 146
Relaxation, 8
Remote area, 15, 78, 80, 86–88, 92, 128, 137, 140, 141, 143, 144, 155, 163, 168, 170, 171, 189, 195
Remote handling, 181, 191, 192, 194
Replaceable cooled plates, 191
Residual gas, 11, 29, 49, 57, 117, 146, 149
Re-solidification, 37, 190
RF discharge, 181
Rhenium (Re), 188, 190
Ripple loss, 42, 138
Runaway electron, 43, 46, 62, 135

S

Saturation, 99, 104, 115, 120, 121, 148, 174, 178, 179, 191, 195
Saturation concentration, 44, 63, 99, 121, 122, 124, 148, 152, 155, 165, 168, 171, 172, 178–180, 183
Saw tooth activity, 42
Scanning laser, 181
Secondary electron, 8, 49, 61, 62, 110

Self-heating, 9
Self-shaping, 189
Self-sputtering, 77, 91
Sheath potential, 118
SiC, 95
Sievert's law, 119
SIMS depth profile, 151, 179
Simulation, 15, 51, 56, 100, 104, 115, 128, 129, 153, 170
Soft-Xray, 10, 23, 25
Space plane, 39, 41
Spectroscopy, 21, 71, 107, 142
Sputter erosion, 77, 188
Square root flux dependence, 177
316 SS, 64
Stainless steel, 10, 11, 64, 111, 126
Static retention, 121, 122, 124, 127, 153, 154
Steady operation, 95
Steady power load, 41, 42
Stored energy, 43, 101, 133, 135
Strike point, 31, 33, 141, 142, 181
Super-permeation, 118, 129
Surface coating, 176
Surface contaminant, 66, 123
Surface melting, 92, 110, 145
Surface oxidation, 182
Surface oxide, 56, 64, 66
Surface recession, 100, 119, 153, 155, 182
Surface saturation, 115, 124, 180
Swelling, 105, 188, 190

T

T accountancy, 147
T breeding, 12, 44, 45, 145, 146, 196
Test limiter, 26, 28, 30, 98, 134
TEXTOR, 11, 20, 25, 26, 28–31, 98, 122, 134, 136, 140, 188
TFTR, 12, 21, 34, 45, 68, 80, 81, 83, 85, 86, 99, 101, 140, 166, 188
Thermal conductivity, 41, 52, 79, 84, 95, 187, 188, 191–193
Thermal contact coefficient, 193, 194
Thermal desorption, 127, 181, 182
Thermal Desorption Spectroscopy (TDS), 142, 177, 182
Thermal fatigue, 96, 100, 188
Thermalized fuel particle, 163
Thermal quench, 134, 135
Thermal shock, 100
Thermal shock resistance, 41
Thomson scattering, 23, 28
Throughput, 5, 15, 109, 116, 127, 128, 145, 146, 148, 153, 154, 162
Tile gap, 75, 81, 85, 86, 88, 89, 140, 155, 163, 170, 171, 173, 192
T inventory, 12, 92, 115, 121, 126–128, 138, 141, 144, 148, 156, 162, 174, 176, 180, 183, 191, 199
TMAP 7, 128
Tore-Supra, 140, 156
Toroidal side, 86, 87
Transient power load, 39, 41, 42, 45, 71, 133
Transmuting element, 187
Trapping, 29, 102, 119, 120, 126, 129, 149, 152, 174, 177–180, 182, 183, 188, 190, 192
Trapping site, 44, 104, 121, 174, 179
Tritium imaging, 34
Triton, 6, 138, 168
T safety, 12, 15, 71, 115, 120, 126, 145, 152, 157, 162, 180–183, 196
Twin limiter, 30–32
Type I ELM, 133, 134
Type III ELM, 134

U

Ultraviolet, 17, 27

V

Vacancy, 52, 67, 68, 109, 119, 121, 165
Vacancy cluster, 52, 119
Vacancy loop, 119
Vacuum pump, 11, 146
Vacuum Ultra Violet (VUV), 10, 24–26, 28
Vacuum vessel, 11, 44, 61, 82, 146, 156, 167, 195
Vapor pressure, 61, 91, 95
Vapor shielding, 10, 37, 99, 104, 109, 193, 196
Vertical Displacement Events (VDE), 95, 96
Vibration, 8
Visible light, 17
Volume recombination, 23
VPS-W, 176

W

Wall cleaning, 12
Wall conditioning, 42, 115, 152, 181
Wall pumping, 12, 71, 99, 109, 115, 119, 149, 180
Wall saturation, 44, 99, 100, 140, 155
Water, 11, 39, 42, 44, 61, 64, 66, 95, 97, 117, 121, 149, 151, 156, 157, 190, 191, 195

Water vapor, 56, 149
Welding, 190
WO_3, 190
W-shaped divertor, 23, 75, 76, 87, 90, 91

X
X-ray, 10, 21, 23, 24, 42, 52, 62